Lecture Notes in Mechanical Engineering

Lecture Notes in Mechanical Engineering (LNME) publishes the latest developments in Mechanical Engineering—quickly, informally and with high quality. Original research reported in proceedings and post-proceedings represents the core of LNME. Volumes published in LNME embrace all aspects, subfields and new challenges of mechanical engineering.

To submit a proposal or request further information, please contact the Springer Editor of your location:

Europe, USA, Africa: Leontina Di Cecco at Leontina.dicecco@springer.com
China: Ella Zhang at ella.zhang@springer.com
India: Priya Vyas at priya.vyas@springer.com
Rest of Asia, Australia, New Zealand: Swati Meherishi at swati.meherishi@springer.com

Topics in the series include:

- Engineering Design
- Machinery and Machine Elements
- Mechanical Structures and Stress Analysis
- Automotive Engineering
- Engine Technology
- Aerospace Technology and Astronautics
- Nanotechnology and Microengineering
- Control, Robotics, Mechatronics
- MEMS
- Theoretical and Applied Mechanics
- Dynamical Systems, Control
- Fluid Mechanics
- Engineering Thermodynamics, Heat and Mass Transfer
- Manufacturing Engineering and Smart Manufacturing
- Precision Engineering, Instrumentation, Measurement
- Materials Engineering
- Tribology and Surface Technology

Indexed by SCOPUS, EI Compendex, and INSPEC.

All books published in the series are evaluated by Web of Science for the Conference Proceedings Citation Index (CPCI).

To submit a proposal for a monograph, please check our Springer Tracts in Mechanical Engineering at https://link.springer.com/bookseries/11693.

Giuseppe L. Conte · Olivier Sename
Editors

Proceedings of the 11th International Conference on Mechatronics and Control Engineering

ICMCE 2023

Springer

Editors
Giuseppe L. Conte
Dipartimento di Ingegneria dell'Informazione
Università Politecnica delle Marche
Ancona, Ancona, Italy

Olivier Sename
CNRS, Grenoble INP, GIPSA-Lab
Université Grenoble Alpes
Grenoble, France

ISSN 2195-4356 ISSN 2195-4364 (electronic)
Lecture Notes in Mechanical Engineering
ISBN 978-981-99-6522-9 ISBN 978-981-99-6523-6 (eBook)
https://doi.org/10.1007/978-981-99-6523-6

This Springer imprint is published by the registered company Springer Nature Singapore Pte Ltd.
The registered company address is: 152 Beach Road, #21-01/04 Gateway East, Singapore 189721, Singapore

Paper in this product is recyclable.

Preface

The 11th International Conference on Mechatronics and Control Engineering (ICMCE 2023) was held as a virtual event during January 29–31, 2023. Researchers from all over the world attended the conference from home and saw the live broadcast of the invited and contributed speeches.

Keynote speeches were delivered by Prof. Daniel Quevedo (IEEE Fellow), from Queensland University of Technology, Australia; Prof. Graziano Chesi (IEEE Fellow), from the University of Hong Kong, Hong Kong, China; Prof. Maciej Michalek, from Poznan University of Technology (PUT), Poland; Prof. Konstantinos Kyriakopoulos (IEEE Fellow), from National Technical University of Athens, Greece; Prof. Elena Zattoni, from Università di Bologna, Italy; and Prof. Sunil K. Agrawal, from Columbia University, USA. On behalf of the organizing committee of the conference, I would like to thank all the keynote speakers for having accepted the invitation. Their talks addressed problems in the area of mechatronics and control engineering that have a large impact on the organization of our modern society and on the use of resources in industrial production.

The conference proceedings of ICMCE 2023 included 12 full papers and they focused on 4 major topics, which are Mechatronics and Control, Control Theory and System Model, Artificial Intelligence Technology in Intelligent Systems, and Mechanical System Design and Maintenance.

The success of the conference is due to the high scientific level of all the contributions as well as to the enthusiasm of all the attendees. The Technical Committee members did a precious job in selecting the papers that were accepted for presentation and for inclusion in this book of proceedings among those that were initially submitted.

Besides the authors, the attendees and the TC members, I would like to thank the members of the International Advisory Committee Pierre Borne, Graziano Chesi and Daniel Quevedo, the Conference Co-Chair Olivier Sename, the Program Chairs Elena Zattoni, Ricardo Ambrocio Ramirez-Mendoza and Mohammad Salah, and the Publicity Chairs Ke-Lin Du, Sergei Alexandrov and Mark Jackson whose work was fundamental for assuring the smooth running of the conference.

Finally, I would like to thank the Conference Organizer for their efforts in assuring the success of ICMCE in the last 11 years and for giving me the opportunity to chair this conference. We truly hope all the participants have found the discussions fruitful and have enjoyed the opportunity for establishing future collaborations.

Ancona, Italy
January 2023

Giuseppe L. Conte
Conference General Chair

Contents

Three-Dimensional Finite Element Analysis and Experiment of Temperature Rise of Permanent Magnet Linear Synchronous Motor

Jianjian Fan and Mengjie Shen

Abstract Permanent magnet linear synchronous motors (PMLSM) are widely used in semiconductor processing, laser cutting, inspection and precision positioning because of high thrust density, high power density, high acceleration, fast response, good reliability and high precision. The torque of the PMLSM is proportional to the input current, that is, to output large electromagnetic thrust, the PMLSM needs a large input current. The large current will cause large losses, mainly copper loss in the winding, which will eventually heat up the winding and other parts and cause temperature rise. High temperature rise will reduce the motor's insulation life or even destroy the insulation, which will reduce the performance of the motor or cause the motor failure. Thus, by studying and analyzing the temperature rise of PMLSM, it is beneficial to achieve rational design or motor operating condition prediction. In this paper, finite element modeling analysis of the three-dimensional temperature field of the motor is conducted based on a linear motor test rig, which includes a linear motor, a sliding table, a linear bearing, a linear module base and a marble platform. The finite element analysis results are compared with the experiment results and they are consistent. This paper provides a reference method for linear motor temperature rise analysis.

Keywords Permanent magnet linear synchronous motor · Temperature field · Finite element · Fluent · Temperature rise experiment

---

J. Fan (✉)
Zhejiang University of Water Resources and Electric Power, Hangzhou, China
e-mail: 26616742@qq.com

M. Shen
Zhejiang Hiatran Electric Technology Co., Ltd, Hangzhou, China
e-mail: shenmengjie@hiatran.com

G. L. Conte and O. Sename (eds.), *Proceedings of the 11th International Conference on Mechatronics and Control Engineering*, Lecture Notes in Mechanical Engineering,
https://doi.org/10.1007/978-981-99-6523-6_1

1 Introduction

Permanent magnet linear synchronous motors (PMLSM) are widely used in semiconductor processing, laser cutting, inspection and precision positioning because of high thrust density, high power density, high acceleration, fast response, good reliability and high precision [1, 2]. The torque of the PMLSM is proportional to the input current, that is, to output large electromagnetic thrust, the PMLSM needs a large input current. The large current will cause large losses, mainly copper loss in the winding, which will eventually heat up the winding and other parts and cause temperature rise. High temperature rise will reduce the motor's insulation life or even destroy the insulation, which will reduce the performance of the motor or cause the motor failure. Thus, by studying and analyzing the temperature rise of PMLSM, it is beneficial to achieve rational design or motor operating condition prediction. However, the influence of the linear module base and test platform of the linear bearing on the temperature rise of the motor has not been taken into account yet.

Due to the special structure and special operating conditions (such as round-trip, transient high thrust, intermittent operating mode and etc.) of linear motors, the thermal analysis of liner motors by thermal resistance-network analysis (TRA) and finite-element analysis (FEA) are significantly different than those conventional rotating motors, as a result, the study of linear motor temperature fields are few. However, as the linear motors are more and more widely used, the related temperature field analysis has been increasingly attention.

A detailed analysis of the temperature field of a water-cooled large-thrust permanent magnet synchronous motors was carried out by FEA, TRA and experiments in [1], and the simulation result and calculated values were consistent with the experiment results. However, this study only performed a two-dimensional analysis, and the analysis report was only the motor itself. In [3], the three-dimensional temperature field of an asynchronous starting permanent magnet motor has been studied. The analysis uses fluent, a FEA software, to model the temperature field and considers the rotor rotation through the equivalent air gap length, while also considering the effect of bearing thermal conduction on the temperature rise.

The temperature rise of a linear motor was investigated by TRA [4], and compared with experiment results the method is proved to be feasible, but TRA requires a lot tedious equivalent calculations for each part of the motor thermal circuit. In [5], the temperature field analysis of a PMLSM which mounted on a movable substrate base plate operating at low speed was carried out by FEA.

In this paper, finite element modeling analysis of the three-dimensional temperature field of the motor is conducted based on a linear motor test rig, which includes a linear motor, a sliding table, a linear bearing, a linear module base and a marble platform. The three-dimensional finite element temperature field model of the motor is firstly established, the losses of the motor were analyzed and the boundary conditions of the temperature field are determined to solve the temperature field. Secondly, temperature rise experiments were conducted based on the prototype. Finally, the simulation and experiment results are compared and they are consistent. This paper provides a reference for the linear motor temperature rise analysis.

2 Motor Temperature Model

The linear motor test system, as shown in Fig. 1, consists of a marble platform, a linear module base, the stator and mover of the PMLSM, the test bench, a connecting device for test and a load liner motor. PMLSM is a single-sided flat type with iron core structure, the stator consists of back iron, permanent magnets, and permanent magnets stainless steel sheath; the mover consists of winding, iron core, slot insulation and epoxy resin. The motor parameters are shown in Table 1. The motor is naturally cooled and the operating speed is 0.2 m/s.

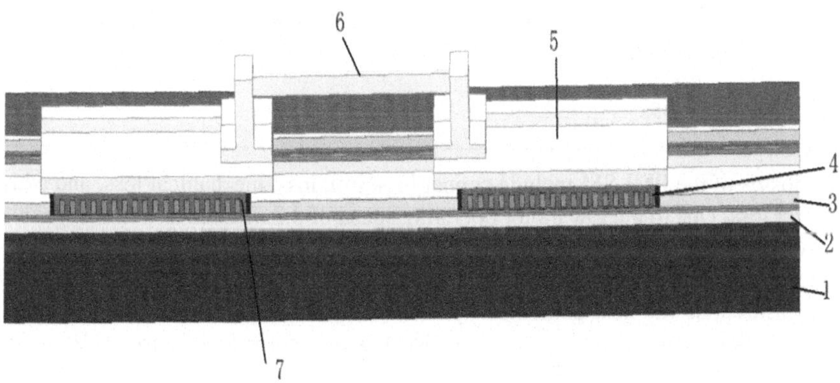

Fig. 1 Temperature rise test platform for PMLSM. 1-Marble platform, 2-Linear module base (linear bearing, slider and guide rail), 3-Stator of linear motor (include back iron, permanent magnets, permanent magnets stainless steel sheath), 4-Mover of the Linear motor, 5-Sliding table, 6-Connecting device for test, 7-Load linear motor

Table 1 Parameters of PMLSM and test rig

Parameters	Value	Parameters	Value
Mover length (mm)	300	Sliding table length (mm)	360
Mover width (mm)	55	Sliding table width (mm)	200
Mover height (mm)	31	Sliding table height (mm)	23.4
Air-gap length (mm)	0.6	Current (A)	4.5
Stator height (mm)	8.4	Line resistance (25 °C) (Ω)	4.5
Module base length (mm)	2500	Marble platform width (mm)	400
Module base width (mm)	210	Marble platform thickness (mm)	130
Module base thickness (mm)	36		

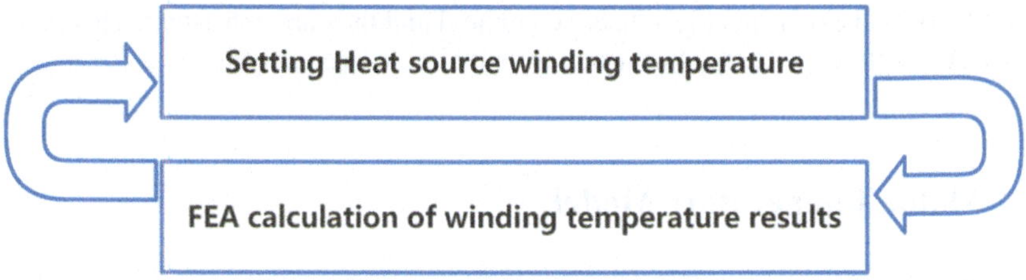

Fig. 2 Temperature iteration of PMLSM

Table 2 Losses of linear motor (25 °C)

Parameters	Value (W)	Percent (%)
Iron loss	0.55	0.39
Eddy current loss	0.01	0.01
Mechanical loss	3.08	2.19
Copper loss	136.68	97.41

2.1 *Motor Loss*

The losses of the PMLSM include copper loss, iron loss, mechanical loss, and eddy current loss. Due to the low motor operating speed of 0.2 m/s, the iron loss and eddy current loss can be neglected. The mechanical loss is mainly the linear bearing loss, which can also be ignored. The friction force of the sliding table is only 15.4 N as measured, so the heat generated by it can be ignored too. The main loss of the motor is the copper loss, $P_{cu} = 1.5*I^2*Rt$, I is the motor RMS current, Rt is the winding resistance at t°C. When the motor winding temperature is 25 °C, the winding resistance is 4.5 Ω, and winding copper loss is 137 W at the rated current. The winding copper loss at t°C can be expressed as $P_{cu} = 137*(235 + t)/(235 + 25)$, and the value of t can be determined by iterative temperature solution, and the iterative process is shown in Fig. 2. The loss value of each part of the motor (mechanical loss is determined by measurement) at 25 °C are shown in the following Table 2. Table 2 illustrates that the other losses in motor can be ignored, so only the copper losses are considered in the analysis.

2.2 *Temperature Field Model*

Given by thermodynamics and [6], In the Cartesian coordinate system, the three-dimensional transient heat conduction equation and its boundary condition is given by

$$\begin{aligned}
&\frac{\partial}{\partial x}\left(k_x\frac{\partial T}{\partial x}\right)+\frac{\partial}{\partial y}\left(k_y\frac{\partial T}{\partial y}\right)+\frac{\partial}{\partial z}(k_z)+q=\rho c\frac{\partial T}{\partial t}\\
&S_1:T=T_0\\
&S_2:k\frac{\partial T}{\partial t}=-q_0\\
&S_3:k\frac{\partial T}{\partial n}=-\mathrm{h}(T-T_e)
\end{aligned} \tag{1}$$

where T is the object temperature, T_0 is the known temperature distribution on the boundary; T_e is the temperature of the surrounding medium; k_x, k_y, k_z are the thermal conductivity of the object in the x, y, z direction, respectively; q is the density of the heat source; q_0 is the density of heat flow through the boundary surface S_2; n is the boundary normal vector; h is the heat transfer coefficient; k is the thermal conductivity; ρ is the density; c is the specific heat.

In order to improve the efficiency of the temperature field finite element model calculation, the following assumptions are made.

1. The copper losses are uniformly distributed by volume in the straight and end sections of the coil.
2. The mover is stationary during the analysis, and the air moves at 0.2m/s relative to the mover.
3. Marble platform temperature is 25 °C, ignore the contact thermal resistance between the linear module base and marble platform. Linear bearing using heat transfer coefficient instead of its actual ball thermal conductivity and lubricant convection heat dissipation, ignore the marble platform and module base.
4. Ignore the temperature conduction between the motor and the load motor. The temperature field model is established as a symmetric half model along the direction of motion.
5. Ignore the openings and chamfer on each component, do not consider the influence of lubrication and thermal radiation on heat transfer, convection heat transfer coefficient is a fixed value, only the copper loss varies with temperature to simplify the modeling and calculation.
6. Separation of copper and insulating varnish.

According to the assumptions and temperature field equations to establish a three-dimensional finite element analysis model of the linear motor as shown in Fig. 3.

The individual material properties in the model are shown in Table 3.

The key to the temperature field solution is to determine the heat transfer coefficient of each face, such as forced convection heat transfer coefficient of each face of the mover, and the natural heat transfer coefficient of each face of the stator. The difficulty in solving the temperature field is to determine the thermal conductivity of the bearing, which involves fluid convection and solid thermal conductivity. The analysis of the heat transfer coefficient involves the following common constants.

Fig. 3 Finite element analysis model of PMLSM

Table 3 Material thermal parameters

Materials	Density (kg/m^3)	Specific heat capacity (k/j)	Thermal conductivity (W/k.m)
Silicon steel sheet	7700	434	X: 39, Y: 4.43, Z: 39
Copper	8890	380	X: 385, Y: 1.5, X: 1.5
Epoxy resin	1800	900	0.8
Aluminum	2700	460	385
Air	1.205	1005	0.026
Slot insulation	930	1340	0.18

Reynolds number:

$$R_e = \frac{ul}{\upsilon} \tag{2}$$

Rayleigh number:

$$R_a = G_r \cdot P_r \tag{3}$$

Grashof number:

$$G_r = \frac{gl^3 \alpha \Delta t}{\upsilon^2} \tag{4}$$

Prandtl number:

$$P_r = \frac{\upsilon}{a} \tag{5}$$

Nusselt number:

$$N_u = \frac{hl}{\lambda} \tag{6}$$

For the convective heat transfer coefficient [7] on the air gap surface is calculated as

For laminar flow ($R_e <= 5 \times 10^5$):

$$N_u = 0.644R_e^{1/2}P_r^{1/3} \tag{7}$$

For turbulent flow ($5 \times 10^5 <= Re <= 10^8$):

$$N_u = \left(0.037R_e^{0.8} - 870\right)P_r^{1/3} \tag{8}$$

For the convective heat transfer coefficient [8] at the front and rear ends of the movers is calculated

$$N_u = 0{,}2R_o^{2/3} \tag{9}$$

For the stator surface, the natural heat transfer coefficient [8] is calculated:

For the horizontal plane:

(a) Upper surface of hot plate or lower surface of cold plate

$$N_u = 0.54R_a^{1/4} \quad (10^4 <= R_a <= 10^7) \tag{10}$$

$$N_u = 0.15R_a^{1/3} \quad (10^7 <= R_a <= 10^{11}) \tag{11}$$

(b) The lower surface of the hot plate or the upper surface of the cold plate

$$N_u = 0.27R_a^{1/4} \quad (10^5 <= R_a <= 10^{10}) \tag{12}$$

For vertical surfaces:

$$N_u = \left\{0.825 + \frac{0.387R_a^{1/6}}{[1 + (0.492/P_r)^{9/16}]^{8/27}}\right\}^2 \tag{13}$$

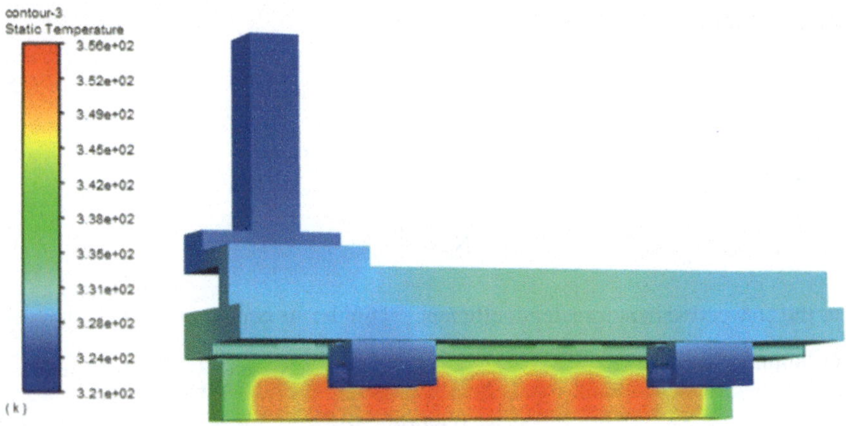

Fig. 4 Steady temperature cloud map of PMLSM

The simulation temperature of the core and winding of the motor is recorded and the temperature curve is shown in Fig. 7. The steady temperature cloud map of the PMLSM is shown in Fig. 4.

3 Experimental Validation

In order to verify the simulation result, the temperature rise test were conducted on the PMLSM as shown in Fig. 5. The temperature thermocouple probe was inserted into the stator core mounting hole and the average temperature rise of the winding was measured by the resistance method. During the test, the operating speed was 0.2 m/s and the distance was 1 m. The motor run in electric mode, while the load motor run in braking mode. The currents of the two motors during the operating cycle were 4.54 and 4.24 Arms. The current curves of the motor are shown in Fig. 5, and the oscilloscope showed a current value of 4.542 Arms during a round trip cycle. The temperature rise of the stator core which measured by the temperature probe is shown in Fig. 6. The line resistance and ambient temperature before and after the test is shown in Table 4.

4 Analysis and Discussion

The FEA results and temperature rise test data are summarized as shown in Table 5, Figs. 7 and 8.

As shown in Table 5, Figs. 7 and 8, it can be seen that the deviation between the finite element results and the measured results is small. The reason for the deviation

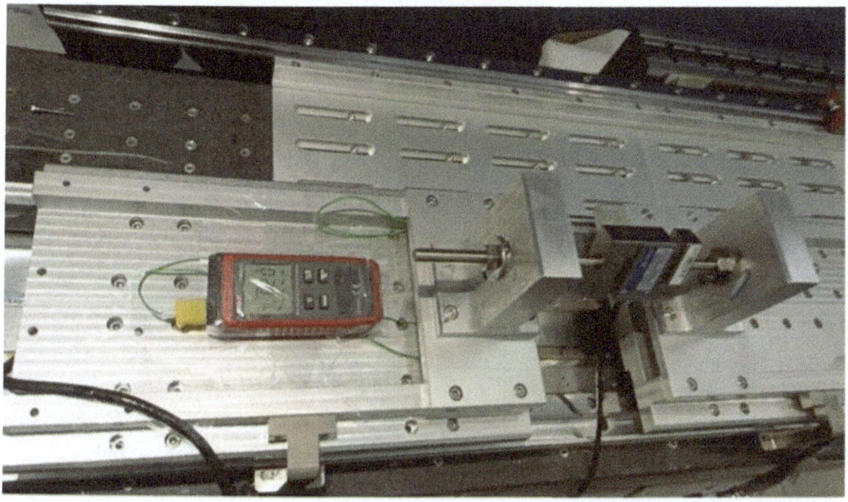

Fig. 5 Temperature rise test for PMLSM

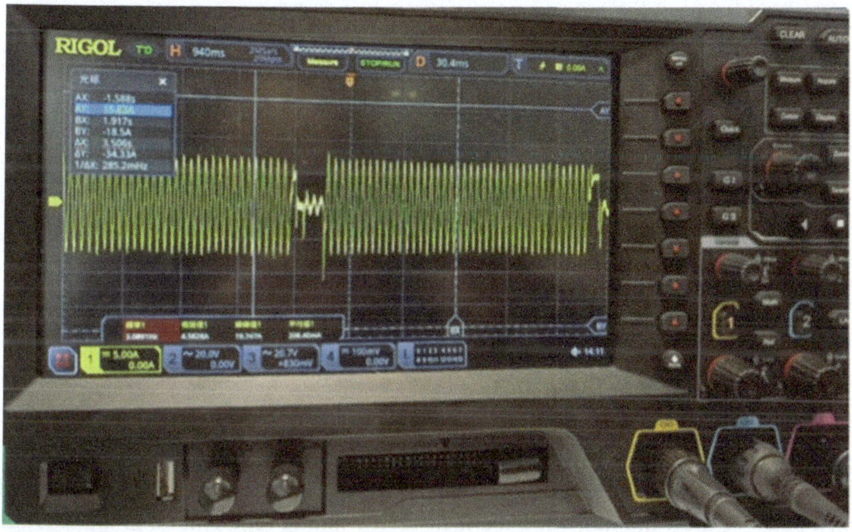

Fig. 6 Current waveform

Table 4 PMLSM temperature rise test winding resistance and temperature

Parameters	Value	Unit
Ambient temperature	25	°C
Line resistance at ambient temperature	4.5	Ω
Line resistance at steady state	5.34	Ω
Average winding temperature at steady state	73.9	°C

Table 5 Comparison of FEA and temperature rise test of PMLSM

Parameters	FEA	Test	Deviation
Winding temperature at steady state (°C)	84.7	73.9	10.8
Stator core temperature at steady state (°C)	62.3	57.1	5.2

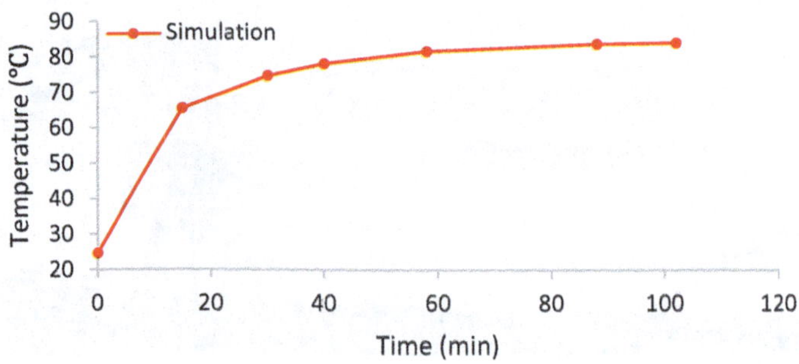

Fig. 7 PMLSM winding temperature rise curve

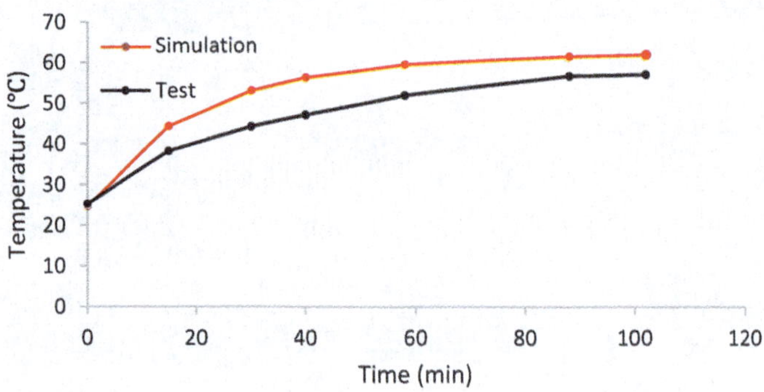

Fig. 8 PMLSM stator core temperature rise curve

of the measurement: the simulation winding temperature is the real-time value at each place of the winding, while the measured winding temperature is the average value [9]. In addition, the measured temperature of the stator core is on the outer surface, which is lower than the true temperature of the core. In the simulation, the heat transfer coefficient calculation also has deviations [10]. In summary, the finite element simulation results higher than the measured temperature, is caused by the existence of measurement errors, and is acceptable.

5 Conclusion

In this paper, based on the PMLSM test platform, a three-dimensional temperature field FEA model of the linear motor is established, and the analysis results are consistent with the experiment results, which verifies the feasibility of the 3D FEA method to analyse the temperature rise of linear motors. Moreover, through the three-dimensional finite element analysis method, the temperature of each part of the linear motor can be more accurately and comprehensively understood, which will help to achieve a more reasonable design.

References

1. Lu Q, Zhang X, et al (2015) Modeling and investigation of thermal characteristics of a water-cooled permanent-magnet linear motor. IEEE Trans Ind Appl:2086–2096
2. Tang Y (2014) Magnetic field analysis and research on electromagnetic force of permanent magnet linear synchronous motor for precision motion platform. Harbin Institute of Technology
3. Li Y, Yan J, Xia J (2015) Temperature field analysis of line-start asynchronous permanent magnet motor based on fluent. J Electr Eng:15–21
4. Amoros JG, Andrada P, Blanque B (2010) An analytical approach to the thermal design of a double-sided linear switched reluctance motor. In: International conference on electrical machines
5. Mu X (2012) Research on the thermal characteristic of a flat permanent-magnet liner synchronous motor feeding system. Zhejiang University
6. Tai Y, Liu Z (2010) Analysis on three-dimensional transient temperature field of induction motor. Proc CSEE:114–120
7. Zhang X, Ren Z, Mei F (2007) Heat transfer, 5th edn. China Construction Industry Press, Beijing
8. Kreith F, et al (2011) Principles of heart transfer.7th edn. Cengage Learning
9. Zhang X (2014) Research on thermal behavior and thrust force of water-cooled permanent magnet liner motor. Zhejiang University
10. Wang H (2014) Fluid dynamics as I understand it. Defense Industry Press, Beijing

Modeling and Control of BLDC Motor for Scaled Autonomous Vehicle Application

Mohamad Hachem, Ariel Medero, Hussam Atoui, and Olivier Sename

Abstract This paper includes a brief study of a BLDC motor considering it an important actuator in new autonomous applications. It can be a sample in treating such actuators from a low level till reaching a controlled closed-loop system. The BLDC motor with the presence of ESC (Electronic Speed Controller) is modeled as a DC motor. Experimental results with gray-box identification techniques were used to validate this assumption. Robust $\mathcal{H}_\infty$ controller is implemented and compared with the conventional PI controller. To improve the controlled system loop an observer is designed and implemented in real-time. It works on the fusion of multiple sensors available on the platform. The targeted actuator is identified and controlled as a part of controlling and developing remote-control (RC) vehicles found in GiPSA-LAB. The main contribution of this paper is to propose a strategy that helps treat the BLDC motors used in scaled autonomous vehicles and pave the way for designing vehicle longitudinal controllers.

Keywords BLDC motors · $\mathcal{H}_\infty$ controller · System identification · Sensor fusion observer

This work has been supported by GiPSA-LAB, safe team. Special thanks to the engineering team in the LAB.

M. Hachem (✉) · A. Medero · H. Atoui · O. Sename
Université Grenoble Alpes, CNRS, Grenoble INP, GIPSA-Lab, 38000 Grenoble, France
e-mail: mohamad.hachem@enac.fr

A. Medero
e-mail: Ariel.Medero@gipsa-lab.grenoble-inp.fr

H. Atoui
e-mail: hussam.atoui@valeo.com

O. Sename
e-mail: olivier.sename@grenoble-inp.fr

H. Atoui
Valeo, Driving Assistance Research (DAR) Team, 94000 Créteil, France

A. Medero
Institut de Robòtica i Informàtica Industrial, C/. Llorens i Artigas 4-6, 08028 Barcelona, Spain

G. L. Conte and O. Sename (eds.), *Proceedings of the 11th International Conference on Mechatronics and Control Engineering*, Lecture Notes in Mechanical Engineering,
https://doi.org/10.1007/978-981-99-6523-6_2

1 Introduction

As one of the most important mechatronics components, DC motors come to the scene of interest of researchers as it plays an important role in the dynamic system. DC motors can be used as main components in applications related to autonomous vehicles [1, 2], autonomous aerial vehicles [3], industrial robotics [4], and other electro-mechanical applications. These applications consider the motor as the main actuator that affects the behavior of the system. Mainly the actuators of the mentioned systems are related to BLDC motor types.

Based on what is mentioned, designing a good controller is a challenge in this field. BLDC motors can be affected by measurement noises and manufacturing uncertainty. To obtain a closed loop, that can track desired references, it is obvious that we need a controller to guarantee system stability and required performances. A fuzzy PID controller is presented by [5] and compare the results with the conventional PID controller. Also [6], proposes a self-tuning fuzzy PID controller targeting autonomous vehicles application. The aim of it is to control the speed of the tire of an electric vehicle based on Pacejka's model. A state of art on different control techniques is presented by [7] and presents modeling techniques and all mechanical and electrical components of BLDC motors. In [3] an adaptive closed-loop controller is used to control actuators of aerial vehicles. The supposed algorithm is proved experimentally to be robust in the scene of parameter uncertainties.

The work presented in this paper forms part of a larger project set to build a full autonomous scaled vehicle platform. The vehicle contains three main actuators, two BLDC motors actuate the longitudinal dynamics of the vehicle while a servo motor acts on the lateral dynamics. The angular speed of the motor is transferred to the angular speed of the vehicle's tire, and then it can be translated to transnational velocity depending on the radius of the tire. The envisioned full autonomous scheme is represented in Fig. 1, shown in green the actuators targeted by this paper. All implementation algorithms and experimental tests were done in the ROS2 framework.

The paper presents a robust controller for the used BLDC motors. The used platform is presented in the first section. Then modeling and identification of the used BLDC are presented in the second section. In the third section a presentation of the theory and development of robust controller based on $\mathcal{H}_\infty$ approach. An observer is designed and implemented in real-time to reduce measurement noises. Experimental results and comparisons between the conventional PI controller and the robust controller are presented in the last section.

2 GiPSA-LAB Platform

The platform found in GiPSA-LAB consists of having the targeted car, a motion-capturing system, and a PC where ROS is running and sending the command input to the Arduino positioned in the vehicle using wireless connections. The full platform

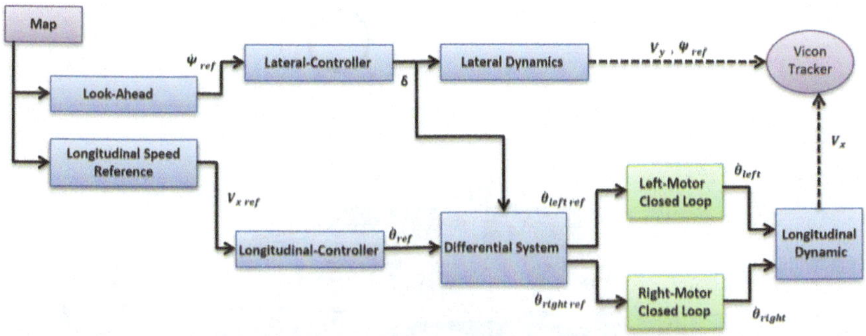

Fig. 1 Vehicle control scheme

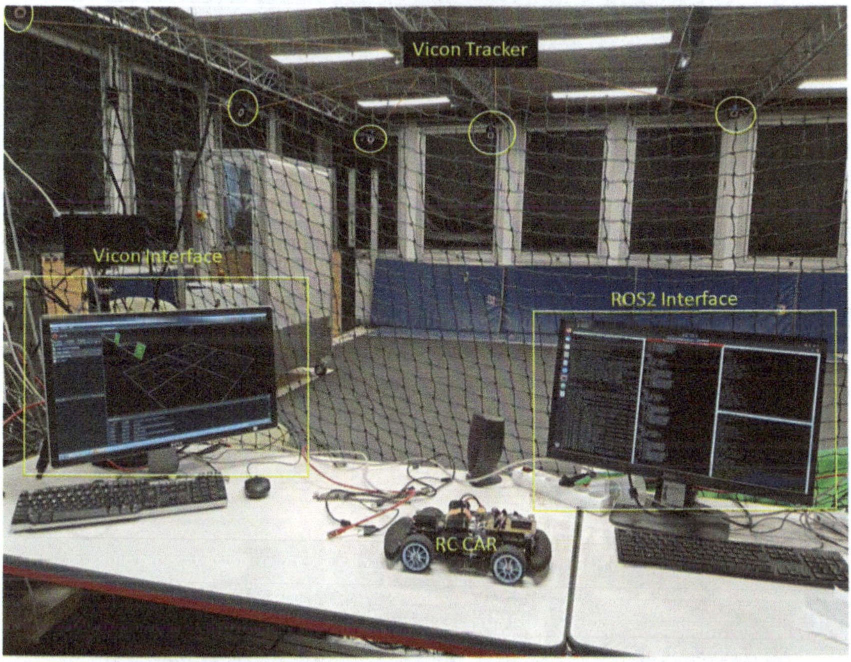

Fig. 2 GiPSA-LAB platform

is shown in Fig. 2. However, it is important to mention that the capturing system (vicon tracker) is not used in the objective of this paper.

The RC car is equipped with two BLDC motors and one servo motor that allow both longitudinal and heading motions, respectively Figs. 3 and 4. This vehicle is equipped with different sensors (battery voltage and current, wheel encoders). The PC runs the control algorithms at specified frequencies and sends PWM signals to the Arduino RP, which transmits the signal as the voltage input to the actuator (Table 1).

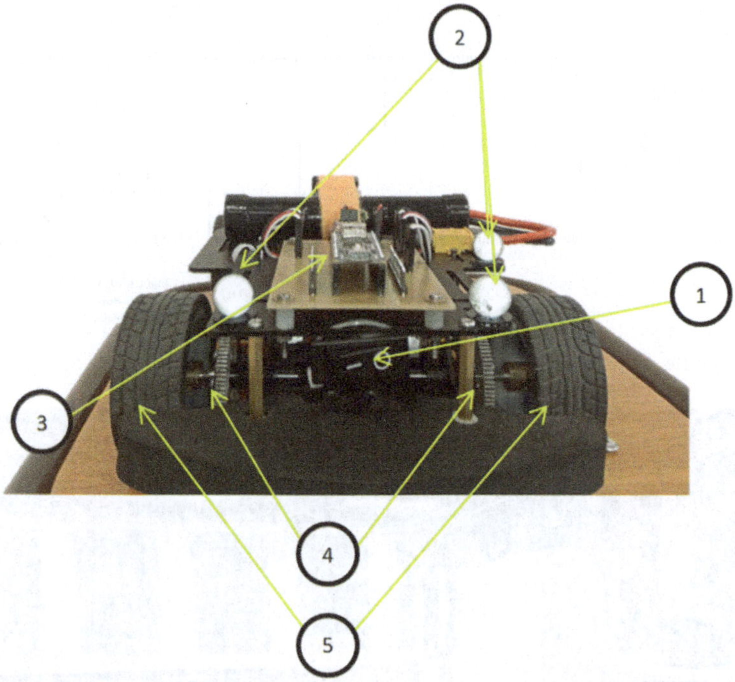

Fig. 3 RC car back view

Fig. 4 RC car front view

Table 1 RC car components

	Type	Functionality
1	Swicth	Switching car power On/OFF
2	8 mm Qualisys super-spherical	Captured by vicon tracker
3	Arduino RP 2040	Micro-controller of the vehicle
4	Spur gears	Increase torque given by BLDC
5	Elastic wheel	2 Rear wheel of the vehicle
6	ACCU NI-MH 3000	Supply power battery
7	MG996R servo motor	Steering actuator
8	Elastic wheel	2 Front wheels of the vehicle
9	BLDC-A2212/13T	Throttle actuator

As mentioned before, the ROS2 framework is used to consist of this development. It is considered an open source that helps researchers and engineers solving complex problems. ROS contains different nodes that can interact and send messages with different structure types using topics. The nodes are running python3 codes at different frequencies. One important thing is containing a ROS bag where all the measured and sent data are collected. This file is treated on Matlab to be able to get clean synchronized data that help in the identification and designing controllers. A joystick is connected also to the platform, and it is used to do different test tasks either by sending a PWM command directly to the Arduino or by generating the desired reference to be tracked by the actuators.

To convert between the input voltage and the PWM signal commanded (1) is used.

$$V = V_{battery} \times PWM \tag{1}$$

Encoder sensors were used to estimate the shaft position of the motor, knowing the time between the signals, it is obvious to estimate the velocity of the output rotor using (2)

$$\dot{\theta}_k = \frac{2\pi}{8 \times Ts}(ttr_k - ttr_{k-1}) \tag{2}$$

where ttr_K represents the number of counted ticks at instant k, T_s corresponds to the working time or frequency of the sensor, and the number 8 is related to the pulse signals read by the sensor at each rotation. It should be mentioned that this velocity estimation produces a noisy signal due to the poor resolution from the encoders. To solve this issue an observer that can estimate a better signal of motor angular velocity is required.

The key system of the scaled autonomous vehicle involved in this work is the actuator component. The BLDC motor used scheme is shown in Fig. 5. We count with a voltage and current sensor for the battery, meaning that the independent

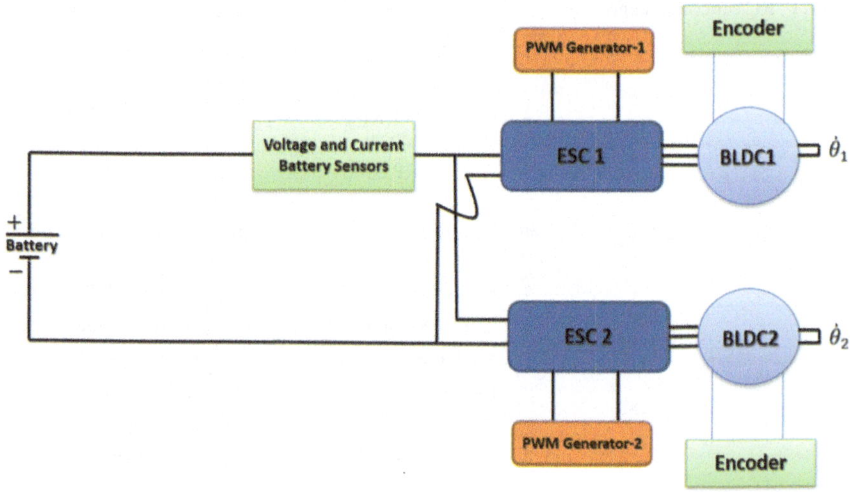

Fig. 5 BLDC schemes

current that each motor consumes is not known. To control each BLDC motor, we count with a dedicated ESC which receives a PWM signal from the Arduino board, controlling then the voltage applied to the motor. The last pair of sensors available, are a hall-effect encoder sensor which is used to derive the angular velocity for each motor.

3 Physical Modeling and System Identification

3.1 Physical Modeling

As mentioned in the previous section, the motor scheme contains an ESC that controls the speed of the motor. However, the system can be modeled as a DC motor from a perspective point of matching input voltage to the output $\dot{\theta}$. The assumed DC scheme is represented in Fig. 6. DC motors consist of having internal inductance coils L, resistors R, inertial rotating rotor J, b motor viscous friction constant, and a stator with magnets to generate an electromagnetic field. Applying Newton's second law and Kirchhoff's voltage law Eq. (3) is derived where $\tau = K_\tau i$, and $e = K_e\dot{\theta}$, with the assumption that $K_\tau = K_e = K$. K_e and K_τ correspond to the emf and motor torque constants.

$$\begin{aligned} J\ddot{\theta} + b\dot{\theta} &= Ki \\ L\frac{di}{dt} + Ri &= V - K\dot{\theta} \end{aligned} \tag{3}$$

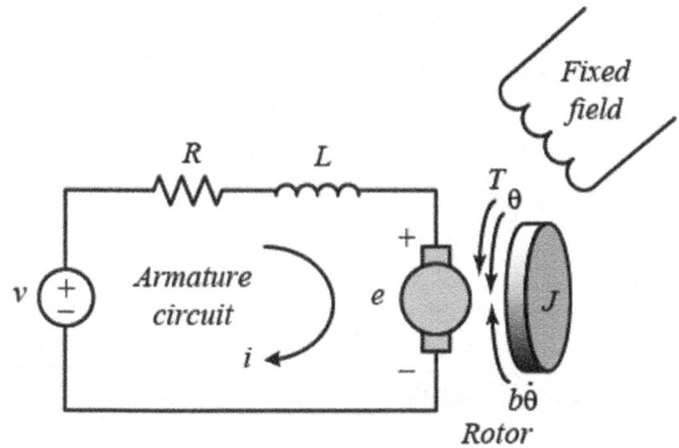

Fig. 6 DC scheme [8]

A linear second-order state-space (4) is constructed based on the results given by (3).Thc input of the the system is voltage V, and the output is the angular speed of the rotor of the motor $\dot{\theta}$. Indeed, the voltage input affects the current i passing through the coil in the motor which affects the rotor angular speed.

$$\frac{d}{dt}\begin{bmatrix}\dot{\theta}\\ i\end{bmatrix}=\begin{bmatrix}-\frac{b}{J} & \frac{K}{J}\\ -\frac{K}{L} & -\frac{R}{L}\end{bmatrix}\begin{bmatrix}\dot{\theta}\\ i\end{bmatrix}+\begin{bmatrix}0\\ \frac{1}{L}\end{bmatrix}V \tag{4}$$

The motor used contains four poles that generate an electromagnetic field that forces the motor rotor to rotate. The ROS2 master can interact with the used microprocessor by sending a PWM signal as a calculated control input.

3.2 *Open Loop Identification*

An important prerequisite for control design is having a good model that can describe the behavior of the system. Benefiting from the mathematical model of the BLDC motor (4) it is possible to undergo parameter estimation for b, K, J, R, L parameters using **Gray-box** identification.

Experimental Setup

An open loop identification test is done on the motor by giving varying PWM amplitude signals. It is worth mentioning that the generated input PWM is considered to be persistently exciting input where it triggers all BLDC targeted frequencies. The signals were generated using the connected JOYSTICK. ROS framework connects

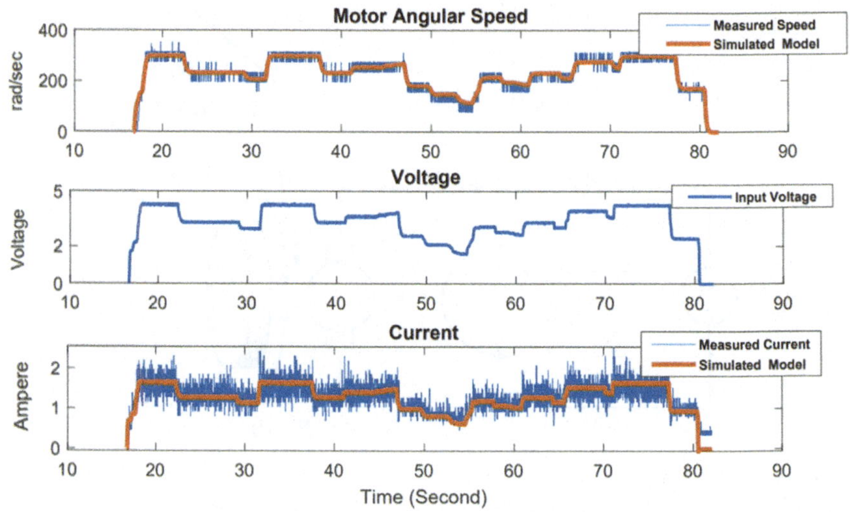

Fig. 7 Motor model identification

the JOYSTICK with the Arduino RP placed on the vehicle and sends it the PWM signals by wireless connection. The measured data can be collected easily as a 'CSV' file thanks to ROSbag and plotjuggler. Plotjuggler is used to plot the collected data after each experiment. It is very important to analyse the collected data set just after finishing each experiment.

Numerical Identification

After treating and filtering the collected data using MATLAB, an optimization problem is used to solve Eq. (5) where y represents the real measured data representing the states of the motor, and $\hat{y}(\hat{\phi})$ represents the simulated states of the model. It is important to mention that while identifying the system constraints on the estimated current were included. This constraints can ensure having feasible solutions while solving an optimization problem. The results of the identification appear in Fig. 7 and Table 2. It is shown that the input delay occurs between 9 and 13 sampling times in the system. The RMSE is calculated to be 0.268.

$$\min_{\hat{\Phi}} \|y - \hat{y}(\hat{\phi})\|_2 \quad \text{Where} \quad \hat{\Phi} = [J, b, K, L, R] \tag{5}$$

3.3 *Validation*

Before designing the controller, a validation of the identified model is performed. Using the previously estimated parameters, and undergoing a new experiment, Fig. 8 represents the validation results. The simulated model fits the real data in both current

Table 2 Motor parameter estimation results

Parameter	Value	S.I Units
J	4.0587e-05	Kg m^2
K	0.0073	V/rad/s
L	1.1201e-04	H
R	2.3923	Ω
b	2.6623e-04	N.m.s

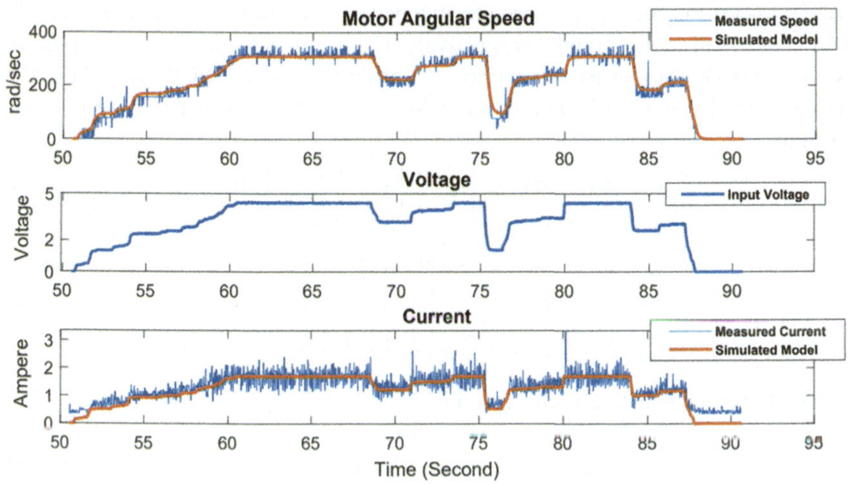

Fig. 8 Motor-estimated model validation

and angular speed measurements. It is worth mentioning that the validation process is done using data sets resulting from different experiments. These sets are collected by either triggering the dynamic behavior of the motor or by trying to perform it stabilized. The represented results are a sample that verifies the model for different objectives.

4 Observer and Control Design

4.1 Sensor Fusion Observer

As seen from Figs. 7 and 8, the available angular speed measurement for the motor speed has a quite high noise to signal ratio due to the low resolution on the motor encoders. However, given that the identified model matches very well the system behaviour, a solution is to use a state observer to filter out the signal noise. One possibility is to have an independent pair of observers for each motor. This design would impose a direct trade-off between confidence on the measurement and confidence on the motor identified model. Another approach is to use the available current measurement coming out from the battery.

As seen in Fig. 5, this sensor can not distinguish the exact current going to each independent motor. However, this measurement can be described mathematically as:

$$i_{bat}(t) = i_1(t) + i_2(t) \tag{6}$$

Were $i_{bat}(t)$ is the current measured for the battery current sensor and $i_{1,2}(t)$ are the current states of motor 1 and 2 respectively. Using this information we propose an extended BLDC Model for the purpose of observer design which fuses the information from the two available motors in the scaled vehicle. This extended model is given by:

$$\frac{d}{dt}\bar{x} = \bar{A}\bar{x} + \bar{B}\bar{V} = \begin{bmatrix} A & 0 \\ 0 & A \end{bmatrix}\bar{x} + \begin{bmatrix} B & 0 \\ 0 & B \end{bmatrix}[V_1,\ V_2]^T \tag{7}$$

$$\bar{y} = \bar{C}\bar{x} = \begin{bmatrix} 1\,0\,0\,0 \\ 0\,0\,1\,0 \\ 0\,1\,0\,1 \end{bmatrix}\bar{x} \tag{8}$$

with the extended state vector given by:

$$\bar{x} = [\dot{\theta}_1\, i_1\, \dot{\theta}_2\, i_2]^T \tag{9}$$

Note that the state matrices A and B in Eq. 7 are those identified from the DC motor model (4). Also note that each motor is assumed to share the same parameter values.

Employing this extended model we make use of a classical Luenberg Observer in order to estimate all states from the motors.

$$\dot{\hat{\bar{x}}}(t) = \bar{A}\hat{\bar{x}}(t) + \bar{B}\bar{V}(t) + L(\bar{y}(t) - \bar{C}\hat{\bar{x}}(t)) \tag{10}$$

The design of the observer gain $L \in \mathbb{R}^{4\times3}$ is accomplished by means of pole placement using Matlab function **place** and using the observer error closed loop dynamics:

$$\dot{e}(t) = (\bar{A} - L\bar{C})e(t) \tag{11}$$

In the rest of the section we describe the control techniques applied for motor control using the information computed by the observer. In contrast to the observer, however, the control of each motor will be carried out independently.

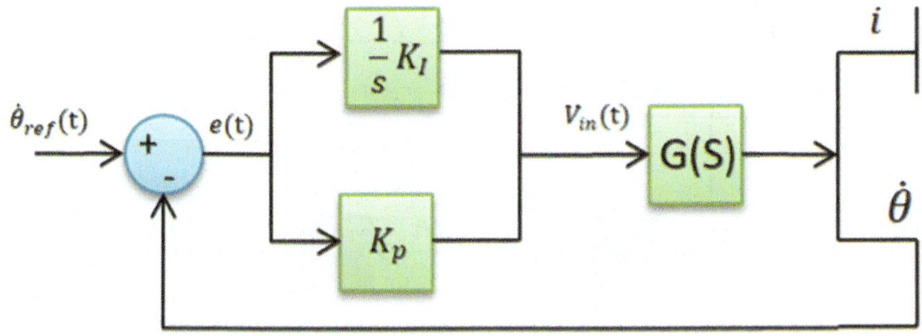

Fig. 9 PI BLDC control simulation architecture

4.2 *PI Controller*

Based on the estimated model of the BLDC motor a PI controller is tuned to control the input voltage of the motor. According to the tuned terms $K_p = 0.16$ and $K_I = 1.1$ can solve the tracking problem. However, since this controller can go to exceed feasible issues, a saturation function is used to bound the input voltage. The used control scheme is represented in Fig. 9.

4.3 $\mathcal{H}_\infty$ *Controller*

An $\mathcal{H}_\infty$ control problem formulation is represented in this part as shown by [9]. The target system to be controlled is represented in Fig. 10 by $G(s)$. Two weighting functions W_e and W_u act as tuning constraints for the built controller $K(s)$. The scheme in Fig. 10 is converted to the scheme LFT shown in Fig. 11.

The generalized plant P aims to characterize the controlled output with the external input and build a controller that minimizes the $\mathcal{L}_2$ induced gain from the external input ω to the controlled output e. In other words, the controller must minimize the impact of the reference and the assumed disturbances and noises with respect to the output of the defined weighting functions. Problem $\mathcal{H}_\infty$ is formulated by (12).

$$\|e\|_2 \leq \gamma\|\omega\|_2 \quad , \quad \|T_{ew}\|_\infty = \left\| \begin{matrix} W_e S \\ W_u K S \end{matrix} \right\| \leq \gamma \tag{12}$$

$$T = \frac{GK}{1 - GK} \quad , \quad S = 1 - \frac{GK}{1 - GK} \quad , \quad SG = S \times G \quad , \quad KS = K \times S \tag{13}$$

S denotes the sensitivity function that indicates the effect of noise on the closed loop T_{yr}, and KS designates the controlled sensitivity function that demonstrates the consequences of external input on controlled input u. It is possible to formulate

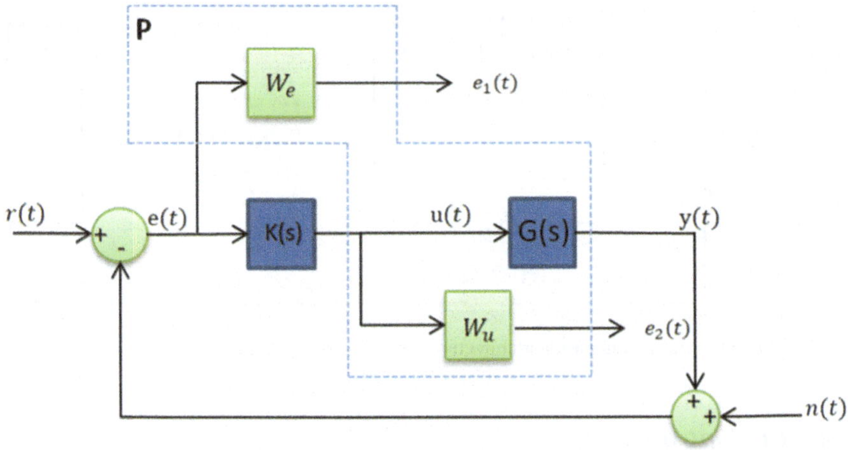

Fig. 10 $\mathcal{H}_\infty$ control architecture

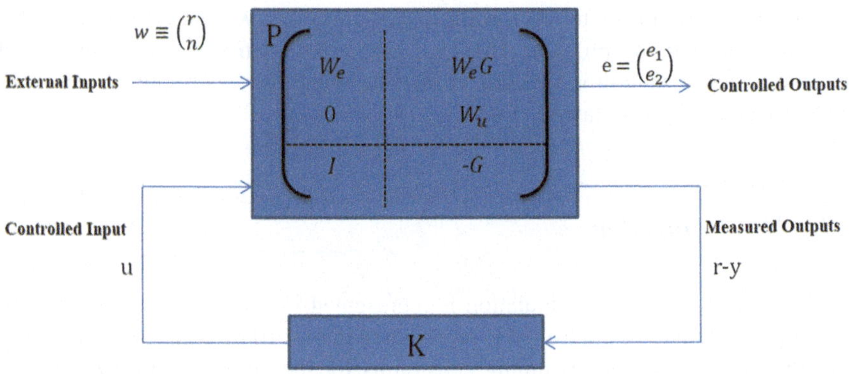

Fig. 11 Generalized plant P

a controller giving $\gamma_{\min}$ regarding the sensitivity functions under defined templates. It is solved by operating the algebraic Riccati approach or using the LMI method (Zhou, Skogestad, & Postlewaite). The weighting functions are defined as W_e and W_u (14) used in designing templates for S and KS, respectively.

$$\frac{1}{W_e} = \frac{s + \omega_b \epsilon_e}{\frac{s}{M_s} + \omega_b} \quad \text{and} \quad \frac{1}{W_u} = \frac{\epsilon_u s + \omega_b C}{s + \frac{\omega_b C}{M_u}} \tag{14}$$

The weighting function is tuned according to the required performance, where:

- M_s: Robustness required with max module margin.
- ω_b: Tracking speed and rejecting disturbances.
- ϵ_e: Steady-state tracking error.

Table 3 BLDC $\mathcal{H}_\infty$ weighting parameters

Parameter	Value
M_s	3.52
ω_b (rad/s)	3.789
ϵ_e	0.00328
M_u	7
ω_{bc} (rad/s)	∞
ϵ_u	10^{-3}

- M_u: Actuator constraints based on $\frac{\Delta u}{\Delta r}$.
- ω_{bc}: Actuator Bandwidth.
- ϵ_u: Attenuate noises on controlled input.

The plant P can be written in another form shown in (15).

$$\begin{aligned} \dot{x} &= Ax + B_1\omega + B_2u \\ e &= C_1x + D_{11}\omega + D_{12}u \\ y &= C_2x + D_{21}\omega + D_{22}u \end{aligned} \tag{15}$$

$x \in \mathbb{R}^n$, the union of the plant states and the weighing function states, $\omega \in \mathbb{R}^{n_\omega}$, the defined external inputs to P, $u \in \mathbb{R}^{n_u}$, the controlled input, $e \in \mathbb{R}^{n_e}$, the controlled outputs, $y \in \mathbb{R}^{n_y}$, the measured outputs of the system. W_e is used to obtain a minimum overshoot and steady-state error, while W_u is used to limit controlled input (V_{in}) and avoid saturation [10]. The weighting function is defined as follows and parameters are defined and tuned according to Table 3 to obtain the required performances.

The generalized plant P of order 3 including a second-order and first-order system. The architecture used to build the controller is represented in (10). The controller $K(s)$ is designed using the Robust Matlab Toolbox. It is important to mention that optimal calculated γ is increased by 30% to ensure the numerical stability of the controller while discretization and implementation. The frequency analysis of the closed-loop system using the $\mathcal{H}_\infty$ approach is shown in Fig. 12. The system is under the defined template in all frequency ranges, has a good steady-state error, and rejects noises at high frequencies.

5 Experimental Implementation

The developed control algorithms have been implemented on the scaled RC car. However, before implementation, different simulation iterations have been done to tune the defined controllers as mentioned previously. In the same step, the designed observer is implemented so that the closed-loop is defined by the feedback of the filtered measured data. It has been implemented with a sampling frequency of 50 Hz.

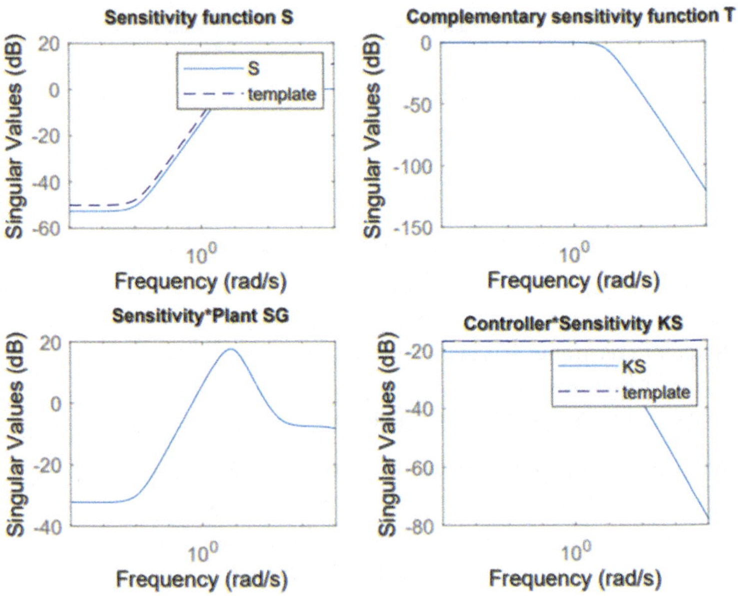

Fig. 12 BLDC closed-loop frequency analysis

As mentioned before the controllers were designed and tuned offline using MATLAB. Python3 programming was used to implement the controllers in real time.

5.1 PI Experimental

Conventional PI control implementation results are shown in Fig. 13. The controlled output appears to follow the reference with a good steady state error thanks to the integrated part of the controller. However, the rising time is large causing slow variation in the dynamics of the controller. This slowness in variation can affect the response of the longitudinal dynamics of the vehicle causing crashes in some maneuvers. The output feedback is filtered using the implemented observer. It is regarded that this controller affects the attenuation of the observer being noisier.

5.2 $\mathcal{H}_\infty$ Control Implementation

Benefiting from the identified model the weighting matrices have been tuned to track the reference. Experimental results are shown in Fig. 14. The $\mathcal{H}_\infty$ controller appears to track the required reference with a small overshoot and steady-state error.

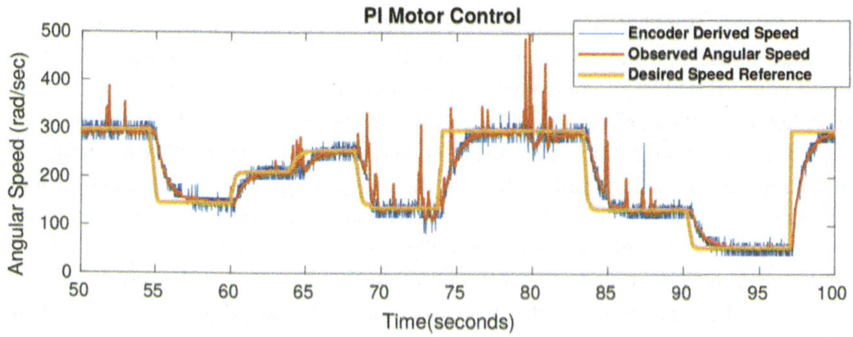

Fig. 13 PI BLDC implementation results

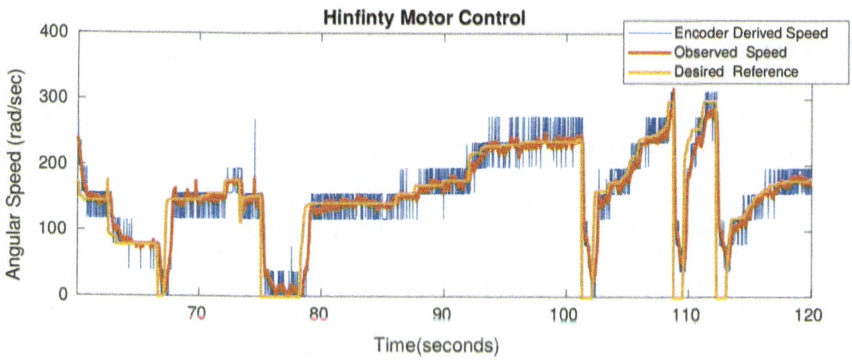

Fig. 14 $\mathcal{H}_\infty$ BLDC implementation results

However, it is obvious that the rising time is small causing fast dynamic variations, thanks to the weighing template "$\frac{1}{We}$". In addition, the used observer appears to be smooth in filtering the noises showing noise attenuation in high frequencies. It is worth mentioning that the controller is not saturating while tracking the desired reference.

6 Conclusion

In this paper, we represent the physical model of a BLDC motor as a DC motor with the presence of ESC. The targeted controller is mainly used in electrical RC cars used to develop autonomous vehicle algorithms. An identification algorithm based on gray-box techniques is used to identify the parameters of the model. The identification method is validated by expiremental data. The data were collected from noisy sensors due to derivation term, however, the method used was able to give a feasible physical value of the motor parameters. PI and $\mathcal{H}_\infty$ control algorithms were introduced and implemented in real-time. A comparison between these two methods shows that the robust $\mathcal{H}_\infty$ controller is better at attenuating noises than the PI controller. The PI controller is easier in tuning and in implementing in real-time, however, the performance of the $\mathcal{H}_\infty$ controller appears to be better in terms of tracking and rising time. Also, the designed observer appears to work greatly in attenuating noises when implemented in the closed loop with the robust controller. The proposed work is validated experimentally and theoretically to be used in treating BLDC motors , specially that used for scaled autonomous vehicle applications. Future work may be comparing different control strategies for the motor with the presence of the longitudinal controller of the vehicle. A study of sensor fault diagnosis can be done in the presence of different type sensors.

References

1. Sayssouk W, Atoui H, Medero A, Sename O. in, 71-85 (2023). ISBN: 978-981-19-1539-0
2. Elakkia E, Anita S, Ganesan RG, Saikiran S (2015) Design and modelling of BLDC motor for automotive applications. Int J Electr Electron Eng Telecommun 1(1):42–48
3. Franchi A, Mallet A (2017) Adaptive closed-loop speed control of BLDC motors with applications to multi-rotor aerial vehicles. In: 2017 IEEE international conference on robotics and automation (ICRA), pp 5203–5208
4. Yen S-H, Tang P-C, Lin Y-C, Lin C-Y (2019) A sensorless and low-gain brushless DC motor controller using a simplified dynamic force compensator for robot arm application. Sensors 19:3171
5. Arulmozhiyal R, Kandiban R (2012) Design of fuzzy PID controller for brushless DC motor. In: 2012 international conference on computer communication and informatics, pp 1–7
6. Baz R, Majdoub KE, Giri F, Taouni A (2022) Self-tuning fuzzy PID speed controller for quarter electric vehicle driven by In-wheel BLDC motor and Pacejka's tire model. IFAC-PapersOnLine 55. In: 14th IFAC workshop on adaptive and learning control systems ALCOS 2022, pp 598–603. ISSN: 2405-8963. https://www.sciencedirect.com/science/article/pii/ S2405896322007789
7. Mohanraj D et al (2022) A review of BLDC motor: state of art, advanced control techniques, and applications. IEEE Access 10:54833–54869
8. Joshi Bharat SRCR (2014) Modeling, simulation and implementation of brushed DC motor speed control using optical incremental encoder feedback
9. Sename O (2020) Robust control of MIMO systems. Ense3 course
10. Krishnan TD, Krishnan CM, Vittal KP (2017) Design of robust H-infinity speed controller for high performance BLDC servo drive. In: 2017 international conference on smart grids, power and advanced control engineering (ICSPACE), pp 37–42. https://doi.org/10.1109/IC-SPACE.2017.8343402.

FedMSPC: A Federated Multivariate Statistical Process Control Framework For Privacy-Preserving Process Modeling Across Company Borders

Du Nguyen Duy, David Gabauer, and Ramin Nikzad-Langerodi

Abstract The ongoing transition from a linear (produce-use-dispose) to a circular economy poses significant challenges to current state-of-the-art information and communication technologies. In particular, the derivation of integrated, high-level views on material, process, and product streams from (real-time) data produced along value chains is challenging for several reasons. Most importantly, sufficiently rich data is often available yet not shared across company borders because of privacy concerns which make it impossible to build integrated process models that capture the interrelations between input materials, process parameters, and key performance indicators along value chains. In the current contribution, we propose a privacy-preserving, federated multivariate statistical process control (FedMSPC) framework based on Federated Principal Component Analysis (PCA) and Secure Aggregation to foster the incentive for closer collaboration of stakeholders along value chains. We tested our approach on two industrial benchmark data sets - SECOM and ST-AWFD. Our empirical results demonstrate the superior fault detection capability of the proposed approach compared to standard, single-party (Multiway) PCA. Furthermore, we showcase the possibility of our framework to provide privacy-preserving fault diagnosis to each data holder in the value chain to underpin the benefits of secure data sharing and federated process modeling.

Keywords Multivariate statistical process control · Federated learning · Privacy-preserving machine learning · Circular economy

D. Nguyen Duy · D. Gabauer · R. Nikzad-Langerodi (✉)
Software Competence Center Hagenberg, Hagenberg, Austria
e-mail: ramin.nikzad-langerodi@scch.at

D. Nguyen Duy
e-mail: du.nguyen.duy@scch.at

D. Gabauer
e-mail: david.gabauer@scch.at

G. L. Conte and O. Sename (eds.), *Proceedings of the 11th International Conference on Mechatronics and Control Engineering*, Lecture Notes in Mechanical Engineering,
https://doi.org/10.1007/978-981-99-6523-6_3

1 Introduction

The importance of data exchange along value chains has been broadly recognized for mastering the transition from linear to circular economy [1, 2]. However, as individuals and corporations are increasingly concerned about how their data are being used, the emphasis on data privacy and security has become a major global topic. There are now data protection obligations that organizations must strictly follow [3]. Therefore, it is challenging, if not impossible, in many situations to transfer data across company borders. This landscape poses new challenges that traditional information and communication technologies in general, and process modeling approaches in particular, cannot handle appropriately.

More specifically, traditional process modeling workflows usually involve collecting and fusing data into a common site where a data-driven model is built. However, this is no longer feasible since data are not owned by a single entity but rather generated and distributed among different companies along a value chain. On the one hand, organizations do not want to share private data because of the fear of trade secrets leaks, and on the other hand due to regulations or geographical restrictions. As a result, even though sufficiently rich process data is available, they often exist in small and fragmented silos and cannot be integrated to enable a broader view of the whole value chain. This is a major obstacle in areas where it is well known that material properties, process parameters, and KPIs are intercorrelated across company borders, for example, in the steel or paper industry [4, 5].

A functional solution to overcome the problem of data fragmentation and isolation is Federated Learning (FL), a concept first proposed by Google in 2016 [6]. The main idea behind FL is to build a centralized model based on data scattered among multiple parties without requiring participants to share sensitive information. In the meantime, FL has gained increasing attention, both from research and industry perspectives. However, a preliminary literature review shows that most of the current work is focused primarily on deep neural networks and their application to computer vision problems [7, 8].

Multivariate Statistical Process Control (MSPC) is an umbrella term for a set of advanced statistical methods for modeling, monitoring, and controlling the operating performance of processes that are widely adopted in the process industry. More specifically, MSPC techniques extract features from high-dimensional and highly correlated process data by means of latent variable (LV) based modeling techniques. The models are then used to monitor processes in real-time, assess their performance, and identify deviations from normal operating conditions (NOC). Therefore, MSPC provides a basis for increasing process security, sustainability, and continuous improvement. Although some recent research efforts have been devoted to adopting federated learning in the field of MSPC, limited progress has been made [9, 10]. Previous studies have focused on federated PCA [11]. However, the application of PCA to MSPC, i.e., for online monitoring, fault detection, and fault diagnosis yet to be proposed. In addition, Multiway PCA (MPCA), an extension of PCA for modeling batch process data has not been investigated, and thus, to the best of our knowledge, the FL paradigm has so far not been adopted for building MSPC-type process models across company borders that preserve the privacy of each contributing party. Moreover, the incentive mechanism, an essential aspect of FL, has not been discussed in the existing literature. A fair value-distribution structure is critical in order to motivate the different parties to actively collaborate in the model training and inference process [12]. The collaboration might be wasteful without meaningful incentives because the participating parties will not carry out efficient contributions.

In this work, we propose a general federated multivariate statistical process modeling framework (FedMSPC) where different companies along a value chain can together build a shared process monitoring model in a federated and privacy-preserving manner. To fully protect confidential data, FedMSPC uses a combination of two privacy techniques: Data Masking and Secure Aggregation [11]. In particular, each participant unfolds and encrypts data in the local environment using a well-designed masking method. Then, all encrypted data are transferred to a third-party server, which securely aggregates these data and trains an MPCA model on the concatenated joint (encrypted) data matrix. Finally, each participant decrypts the federated output of the model, using its private key, and reconstructs the information related to their process. More specifically, each participant obtains only the portion of the shared loadings matrix that corresponds to the variables that they contribute. Thus, this information is not accessible to the other parties. In addition, all participants will share the explained variance corresponding to the selected principal components (PCs). Using such results, all data holders can collaboratively estimate the scores, process-monitoring statistics (e.g. Hotelling's T^2 and Q-statistic) as well as variable contributions, thereupon conduct fault detection and diagnosis.

In order to showcase the feasibility of the framework, we propose Federated Principal Component Analysis (FedPCA), which is based on the idea of Federated Singular Value Decomposition (FedSVD) proposed in [11], as the modeling method. FedSVD basically provides lossless privacy guarantees and is thus ideally suited for building federated MSPC models. However, in [11], the authors employ FedPCA under a horizontally partitioned scenario, where the data from the contributing parties share the same feature rather than the sample space (i.e. horizontal FL). However, in value chains, input materials are processed sequentially by different companies and the corresponding data is thus vertically partitioned, i.e., the data share the same sample space but different feature spaces. Therefore, we will concentrate on vertical FedPCA instead. In addition, we will investigate the application of FedPCA in modeling batch processes.

2 Methodology

2.1 Multivariate Statistical Process Control

PCA is among the best-known and most widely adopted MSPC techniques for modeling high-dimensional and highly correlated process data, and will be introduced in the following.

Principal Component Analysis (PCA) is often applied in MSPC to transform a dataset with highly correlated variables into an uncorrelated dataset while preserving only the systematic variation. There are various techniques for building a PCA model, Singular Vector Decomposition (SVD) being the most popular one. Suppose the original dataset denoted by $\boldsymbol{X} \in \mathbb{R}^{m \times n}$ contains m observations and n process variables. SVD corresponds to the following decomposition:

$$\begin{aligned} \boldsymbol{X} &= \boldsymbol{U}\boldsymbol{\Sigma}\boldsymbol{V}^T \\ &= \left[\boldsymbol{U}_r\ \boldsymbol{U}_0\right]\begin{bmatrix} \boldsymbol{\Sigma}_r & 0 \\ 0 & \boldsymbol{\Sigma}_0 \end{bmatrix}\left[\boldsymbol{V}_r\ \boldsymbol{V}_0\right]^T, \end{aligned} \tag{1}$$

where $\boldsymbol{U} \in \mathbb{R}^{m \times m}$ is the left singular matrix, $\boldsymbol{\Sigma} \in \mathbb{R}^{m \times n}$ is the diagonal matrix holding the singular values, and $\boldsymbol{V} \in \mathbb{R}^{n \times n}$ is the right singular matrix. $\boldsymbol{V}_r \in \mathbb{R}^{n \times r}$ derived from $\boldsymbol{V}$ is called the loadings matrix. The number of principal components r can be determined based on a certain criterion, e.g., the cumulative explained variance, and usually it holds that $r \ll n$. The loadings are the coefficients of the variables from which the PCs are computed. The sign of the loading shows whether the correlation between the PCs and the corresponding variable is positive or negative while its absolute value indicates how strongly the variable influences the PCs. Therefore, they are often used to quantify variable importance.

Projecting $\boldsymbol{X}$ onto the subspace spanned by the selected PCs reduces the dimensionality of the column space from n to r. The result of the transformation is the scores matrix $\boldsymbol{T}_r \in \mathbb{R}^{m \times r}$

$$\boldsymbol{T}_r = \boldsymbol{X}\boldsymbol{V}_r. \tag{2}$$

The reconstruction of $\boldsymbol{X}$ can be estimated from $\boldsymbol{T}_r$ and $\boldsymbol{V}_r$:

$$\hat{\boldsymbol{X}} = \boldsymbol{T}_r\boldsymbol{V}_r^T \tag{3}$$

The residual matrix $\boldsymbol{E}$ is defined as the errors between $\boldsymbol{X}$ and $\hat{\boldsymbol{X}}$ and can be calculated as

$$\boldsymbol{E} = \boldsymbol{X} - \hat{\boldsymbol{X}} = \boldsymbol{X} - \boldsymbol{X}\boldsymbol{V}_r\boldsymbol{V}_r^T. \tag{4}$$

Multi-way principal component analysis (MPCA) is an extension of PCA for monitoring batch processes that are broadly seen in industries where batch and semi-batch process operations are common, for example, in the chemical or pharmaceutical industry [13, 14].

Suppose each batch run has J variables measured at K time intervals. Similar data exist in the same form for each of the I batch runs. Therefore, all the data can be arranged in an array $\boldsymbol{X}(I \times J \times K)$.

In order to apply PCA, the dataset has to be converted into a two-dimensional array. There are multiple ways to unfold a 3D dataset. However, the most meaningful approaches are batch-wise and variable-wise unfolding. In the proposed approach, we will employ batch-wise unfolding since variable-wise unfolding is not feasible for federated MSPC. In batch-wise unfolding, the 2D array is formed by unfolding the array $\boldsymbol{X}$ so that each of its vertical slices contains the observed variables for all batches at a

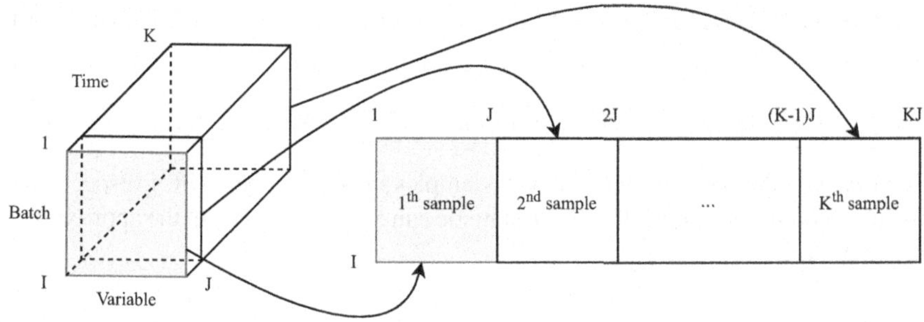

Fig. 1 An illustration of the batch-wise unfolding method

specific time instance. The result is a 2D matrix of the shape $(I \times KJ)$. An illustration of this unfolding technique is shown in Fig. 1.

After the data is unfolded, PCA is performed to retrieve the scores and loadings matrices. Hotelling's T^2, Q-statistics, and the corresponding variable contributions can then be calculated in a similar manner as for PCA.

From an operational point of view, it is preferable to monitor the batch as it progresses in order to anticipate process faults and to take timely actions to prevent out-of-batch-specification events. However, in this situation, a major obstacle is that the new batch is required to have KJ columns such as the NOC data used for training the model. This is impossible when the batch has not been completed because at time interval $k < K$ the new batch x only has kJ columns. In [14], the authors proposed a simple solution to overcome this problem which is to use only the portion of the loadings matrix that corresponds to the elapsed time period until the current time interval k to calculate the new scores vector

$$t_r^{[k]} = x\tilde{\boldsymbol{V}}_r(\tilde{\boldsymbol{V}}_r^T\tilde{\boldsymbol{V}}_r)^{-1}, \tag{5}$$

where $\tilde{\boldsymbol{V}}_r = \boldsymbol{V}_r[1:kJ]$ contains the first kJ columns of $\boldsymbol{V}_r$. In the proposed approach, we will use this same approach to handle the situation where batches are incomplete.

Fault detection and diagnosis One of the most popular applications of PCA-based MSPC is fault detection and diagnosis. This is often done based on control chart statistics, such as Hotelling's T^2 and the so-called Q-statistic. While a high Q-statistic indicates a change in covariance structure, a high T^2 indicates that although the sample is described well by the model (i.e. through a linear combination of the loadings) it is unusual in terms of the linear combination. Suppose there is a PCA model generated from NOC data. When a new sample $x \in \mathbb{R}^{1\times n}$ arrives, T^2 and Q can be calculated as:

$$T^2 = t_r\boldsymbol{\Lambda}_r^{-1}t_r^T, \tag{6}$$

where $t_r = x\boldsymbol{V}_r$ and $\boldsymbol{\Lambda}_r = \boldsymbol{\Sigma}_r^2$ and

$$Q = \sum_{i=1}^{n}(x_i - \hat{x}_i)^2. \tag{7}$$

The upper confidence limit for the T^2 statistic can be computed from the F-distribution

$$T_\alpha^2 = \frac{r(m-1)}{r-m} F_{r,m-r,\alpha}, \tag{8}$$

where m and r denote the number of NOC samples and the number of PCs, respectively. The upper confidence limit for the Q-statistic can be computed from the approximate distribution

$$\begin{aligned} Q_\alpha &= \theta_1 \left(1 - \frac{\theta_2 h_0 (1-h_0)}{\theta_1^2} + \frac{z_\alpha \sqrt{2\theta_2 h_0^2}}{\theta_1} \right)^{1/h_0} \\ \theta_i &= \sum_{j=r+1}^{l} (\mathbf{\Lambda}_{j,j})^i, i = 1, 2, 3 \\ h_0 &= 1 - \frac{2\theta_1\theta_3}{3\theta_2^2}, \end{aligned} \tag{9}$$

where z_α is the standard normal deviate corresponding to the upper $(1-\alpha)$ percentile, $\mathbf{\Lambda}_{jj}$ is the eigenvalue associated with the jth loading vector, and l is the number of non-zero eigenvalues calculated from the data.

A sample might be considered faulty if either T^2 or Q exceeds the predefined control limits T_α^2 or Q_α. When a fault is detected, variable contributions to T^2 and Q can be calculated as [14]

$$T_{cont}^2 = x\mathbf{V}_r\mathbf{\Sigma}^{-1}\mathbf{V}_r^T = t_r\mathbf{\Sigma}^{-1}\mathbf{V}_r^T \tag{10}$$

and

$$Q_{cont} = (x - \hat{x})^2. \tag{11}$$

Variables with high contributions are diagnosed as candidates for the cause of the fault.

2.2 *Federated Multivariate Statistical Process Control (FedMSPC)*

Figure 2 illustrates the proposed framework. Assume we have g data holders with the i-th data holder owning data matrix $\boldsymbol{X}_i \in \mathbb{R}^{m \times n_i}$ and these data holders aim at fitting a PCA model on the concatenated matrix $\boldsymbol{X} = [\boldsymbol{X}_1, \boldsymbol{X}_2, ..., \boldsymbol{X}_g]$, where $\boldsymbol{X} \in \mathbb{R}^{m \times n}$ and $n = \sum_{i=1}^{g} n_i$. In this case, the full results of PCA include $\mathbf{\Sigma} \in \mathbb{R}^{m \times n}$ and $\boldsymbol{V}^T = [\boldsymbol{V}_1^T, ..., \boldsymbol{V}_g^T] \in \mathbb{R}^{n \times n}$, where $\boldsymbol{V}_i^T \in \mathbb{R}^{n \times n_i}$ is the portion of the loadings matrix corresponding to the variables contributed by the i-th data holder.

The aim of this contribution is to design a privacy-preserving system that guarantees that (1) during the computation, data is not leaked to any other data holder and (2) the loadings matrix $\boldsymbol{V}$ must be vertically and secretly distributed among data holders. While the first requirement is commonly shared by all privacy-preserving applications, the second requirement is more specific to federated MSPC. As explained in the previous chapter, loadings can reveal sensitive information about how variables interact with the principal components and can be used to calculate the contributions of variables to T^2 and Q statistics. Therefore, data holder i should know only $\boldsymbol{V}_i \in \mathbb{R}^{n_i \times n}$ that contains coefficients corresponding to its contributed variables. Furthermore, $\boldsymbol{V}_i$ has to be unknown to all other involved parties. In order to realize these goals, we propose an approach based on FedSVD [11].

To apply FedSVD-based PCA as proposed in [11], the system requires a Trusted Authority (TA) to handle key generation, and a Computation Service Provider (CSP) to take care of data aggregation and model building. Algorithm 1 shows the overall model-building workflow. Since $\boldsymbol{\Sigma}$, the number of NOC samples, the number of variables, and the number of principal components are shared by all data holders, the control limits for Hotelling's T^2 and Q-statistic can be estimated using Eqs. 6 and 7 as for standard PCA.

After a FedPCA model is built, suppose a new sample $x = [x_1, ..., x_g]$, where $x_i \in \mathbb{R}^{1 \times n_i}$, is generated. Algorithm 2 is used to calculate scores, Hotelling's T^2 and Q-statistics, and the contribution of each variable to these quantities. Using these monitoring values, all data holders can check whether the sample/batch is faulty or not, i.e. if the monitoring statistics lie above their critical limits. However, since $\boldsymbol{V}_i$ is secretly owned by each data holder, the computation of Hotelling's T^2- and Q-contributions can only be done locally. This might be greatly beneficial since all participating companies might know that a sample/batch is faulty, and they can check whether the problem might be associated with their production line. However, each company only sees the contri-

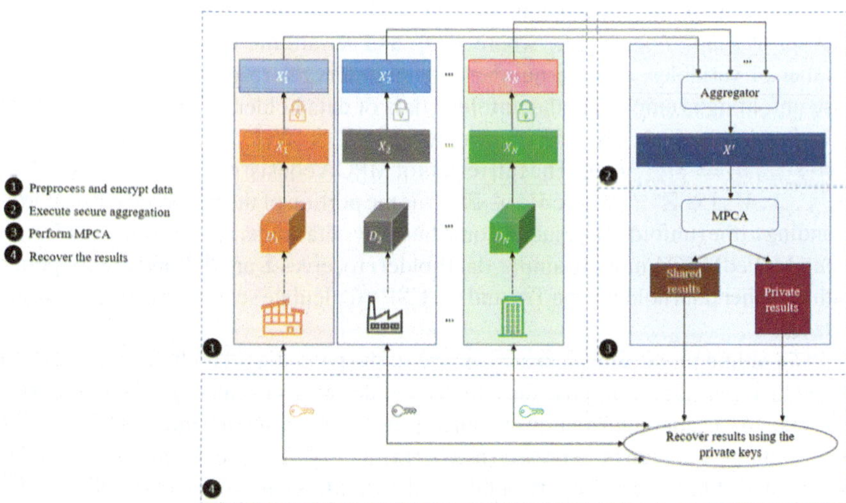

Fig. 2 The architecture of the federated process modeling framework

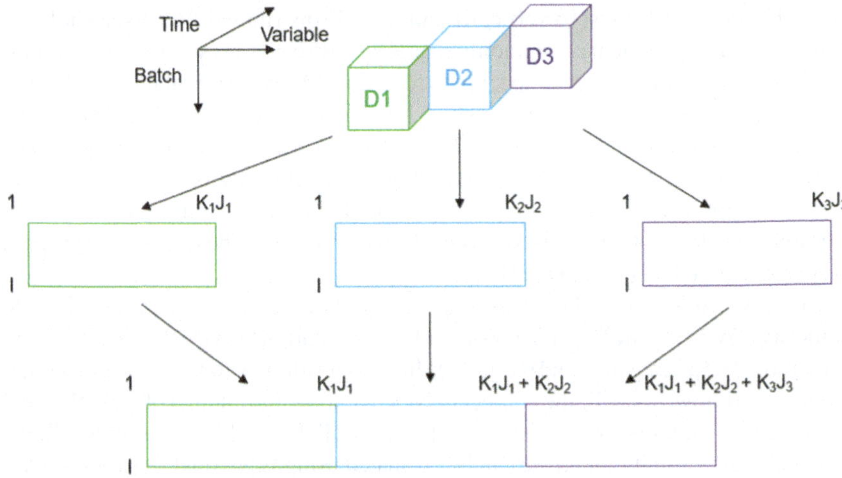

Fig. 3 An illustration of how local data is mapped batch-wise in FedMPCA

bution of its own variables to the fault, which provides feedback on how to improve its own process to benefit the entire value chain.

For batch process data, FedPCA cannot be applied directly. Thus, we propose an extension called FedMPCA that includes a data unfolding step before encryption and transfer to the CSP. However, unlike the transition from PCA to MPCA, not all unfolding techniques are applicable in the federated scenario. While variable-wise unfolding is invalid because local data do not share the same feature space, batch-wise unfolding is undertaken as shown in Fig. 3.

Suppose there are g participating data holders and data holder i owns a batch data set of shape $I \times J_i \times K_i$ where I, J_i, and K_i are the number of batches, the number of variables, and the number of time intervals, respectively. When batch-wise unfolding is employed, the unfolded data of data holder i is $\boldsymbol{X}_i \in \mathbb{R}^{m \times n_i}$ where $m = I$ and $n_i = K_i J_i$. Therefore, the joined data $\boldsymbol{X} = [\boldsymbol{X}_1, \ldots, \boldsymbol{X}_g] \in \mathbb{R}^{m \times n}$, where $n = \sum_{i=1}^{g} n_i = \sum_{i=1}^{g} K_i J_i$. The full results of MPCA consist of $\Sigma \in \mathbb{R}^{m \times n}$ and $\boldsymbol{V}^T = [\boldsymbol{V}_1^T, \ldots, \boldsymbol{V}_g^T] \in \mathbb{R}^{n \times n}$, where $\boldsymbol{V}_i^T \in \mathbb{R}^{n \times n_i}$ is the portion of the loadings matrix corresponding to the (unfolded) variables contributed by data holder i. Similar to FedPCA, at the end of FedMPCA model training, data holder i receives $\boldsymbol{\Sigma}$ and $\boldsymbol{V}_i^T$ and can cooperate with the other data holders, the TA, and the CSP to calculate scores and the monitoring indexes.

In an online monitoring scenario, given all the preceding data holders have completed their processes, suppose the i-th data holder wants to calculate the monitoring indexes and variable contributions for an in-progress batch at time interval $k < K$. The corresponding data can be expressed as $x = [x_1, ..., x_i]$. The procedure is described in Algorithm 3. Once the scores are obtained, data holders can calculate Hotelling's T^2,

Q-statistics and contribution of variables to the two indexes in a similar manner as it is done for completed batches (described in lines 16-37 of Algorithm 2).

Algorithm 1: FedPCA Training

Input: $X = [X_1, \dots, X_g]$
Output: $\boldsymbol{\Sigma}$, r and $V^T = [V_1^T, \dots, V_g^T]$
Constraint: Data holder i's data is not leaked, and it receives $\boldsymbol{\Sigma}$, r and V_i^T as results.

Function `FedPCA.train`($[X_1, \dots, X_g]$)
  TA do:
    Generate orthogonal matrices $P \in \mathbb{R}^{m \times m}$, $B \in \mathbb{R}^{n \times n}$.
    Then split B^T into $[B_1^T, \dots, B_g^T]$ where $B_i^T \in \mathbb{R}^{n \times n_i}$.
  end
  Data Holders do:
    for $i = 1 \rightarrow g$, *Data holder i* **do**
      Download P, B_i^T from TA and compute
      $X_i^{'} = PX_iB_i$
    end
  end
  CSP do:
    Aggregate $X^{'}$:
    $X^{'} = \sum_{i=1}^{g} X_i^{'} \quad (= \sum_{i=1}^{g} PX_iB_i = PXB)$
    Perform standard SVD:
    $X^{'} = U^{'}\Sigma V^{'T}$
  end
  Data Holders do:
    for $i = 1 \rightarrow g$, *Data holder i* **do**
      Download $\boldsymbol{\Sigma}$ from CSP.
      Determine the number of principal components r based on the cumulative sum of explained variance.
      Generate a random matrix $R_i \in \mathbb{R}^{n_i \times n_i}$
      Mask B_i^T through: $[B_i^T]^{R_i} = B_i^T R_i$
      Send $[B_i^T]^{R_i}$ to CSP.
    end
  end
  CSP wait to receive data and do:
    if *Receive* $[B_i^T]^{R_i}$ *then* **then**
      Compute $[V_i^T]^{R_i} = V^{'T}[B_i^T]^{R_i} \quad (= V^{'T}B_i^T R_i = V_i^T R_i)$
      Send $[V_i^T]^{R_i}$ back to data holder i.
    end
  end
  Data Holders do:
    for $i = 1 \rightarrow g$, *Data holder i* **do**
      Receive $[V_i^T]^{R_i}$ from CSP.
      Recover V_i^T by $V_i^T = [V_i^T]^{R_i} R_i^{-1}$.
    end
  end
End

Algorithm 2: FedPCA Inference

Input: $x = [x_1, ..., x_g]$
Output: t_r, Q-statistics, Hotelling's T^2, $Q_{cont} = [Q_{cont,1}, ..., Q_{cont,g}]$,
$T^2_{cont} = [T^2_{cont,1}, ..., T^2_{cont,g}]$
Constraint: Data holder i's data is not leaked, and it receives t_r, Q-statistics, Hotelling's T^2, $Q_{cont,i}$, $T^2_{cont,i}$ as results.

Function `FedPCA.predict` $([x_1, ..., x_g])$
  TA do:
    Generate a random number p.
  end
  Data Holders do:
    for $i = 1 \rightarrow g$, *Data holder i* **do**
      Download the random number p from TA.
      Calculate local scores: $t_{r,i} = x_i \boldsymbol{V}_{r,i}$
      Encrypt the local scores with p: $t'_{r,i} = pt_{r,i} = px_i \boldsymbol{V}_{r,i}$
      Send $t'_{r,i}$ to CSP.
    end
  end
  CSP do:
    Aggregate t'_r: $t'_r = \sum_{i=1}^{g} t'_{r,i} \quad (= p \sum_{i=1}^{g} x_i \boldsymbol{V}_{r,i} = px\boldsymbol{V}_r)$
  end
  Data Holders do:
    for $i = 1 \rightarrow g$, *Data holder i* **do**
      Download t'_r from CSP.
      Recover the scores t_r: $t_r = \frac{t'_r}{p}$
      Calculate T^2: $T^2 = t_r \boldsymbol{\Sigma}^{-2} t_r^T$
      Calculate $T^2_{cont,i} = t_r \boldsymbol{\Sigma}^{-1} \boldsymbol{V}^T_{r,i}$
      Calculate local reconstruction errors e_i: $e_i = x_i - t_r \boldsymbol{V}^T_{r,i}$
      Calculate $Q_{cont,i}$: $Q_{cont,i} = e_i^2$
      Calculate local Q-statistics: $Q_i = e_i e_i^T$
      Encrypt Q_i using p: $Q'_i = pQ_i$
      Send Q'_i to CSP.
    end
  end
  CSP do:
    Aggregate Q': $Q' = \sum_{i=1}^{g} Q'_i \quad (= p \sum_{i=1}^{g} Q_i = pQ)$
  end
  Data Holders do:
    for $i = 1 \rightarrow g$, *Data holder i* **do**
      Downloads Q' from CSP.
      Recover Q by $Q = \frac{Q'}{p}$
    end
  end
End

Algorithm 3: FedMPCA Incomplete Batch

Input: $x = [x_1, ..., x_i]$
Output: $t_r^{[k]}$
Constraint: Data holder i's data is not leaked, and it receives $t_r^{[k]}$ as the result.

Function `FedMPCA.predict`$([x_1, ..., x_i])$
  TA do:
    Generate a random number p and an random matrix $\boldsymbol{W} \in \mathbb{R}^{r \times r}$
  end
  Data Holders do:
    for $j = 1 \rightarrow i$*, Data holder* j **do**
      Download the p and $\boldsymbol{W}$ from TA.
      Calculate $t'_{r,j}$ and $\boldsymbol{F}'_j$ as follows:
      $t'_{r,j} = px_j \tilde{\boldsymbol{V}}_{r,j} \boldsymbol{W}$
      $\boldsymbol{F}'_j = \tilde{\boldsymbol{V}}_{r,j}^T \tilde{\boldsymbol{V}}_{r,j} \boldsymbol{W}$
      where $\tilde{\boldsymbol{V}}_{r,j} = \boldsymbol{V}_{r,j}$ for $j < i$, and $\tilde{\boldsymbol{V}}_{r,j} = \boldsymbol{V}_{r,j}[1 : kJ_j]$ when $j = i$. Note that in this case, $\tilde{\boldsymbol{V}}_r^T = [\tilde{\boldsymbol{V}}_{r,1}^T, ..., \tilde{\boldsymbol{V}}_{r,i}^T]$.
      Send $t'_{r,j}$ and $\boldsymbol{F}'_j$ to CSP.
    end
  end
  CSP do:
    Calculates $t_r^{[k]'}$:

$$t_r^{[k]'} = \sum_{j=1}^{i} px_j \tilde{\boldsymbol{V}}_{r,j} \boldsymbol{W} (\sum_{j=1}^{i} \tilde{\boldsymbol{V}}_{r,j}^T \tilde{\boldsymbol{V}}_{r,j} \boldsymbol{W})^{-1}$$
$$\left(= p \sum_{j=1}^{i} x_j \tilde{\boldsymbol{V}}_{r,j} \boldsymbol{W} \boldsymbol{W}^{-1} (\sum_{j=1}^{i} \tilde{\boldsymbol{V}}_{r,j}^T \tilde{\boldsymbol{V}}_{r,j})^{-1} \right)$$
$$\left(= px \tilde{\boldsymbol{V}}_r (\tilde{\boldsymbol{V}}_r^T \tilde{\boldsymbol{V}}_r)^{-1} \right)$$
$$\left(= pt_r^{[k]} \right)$$

    Broadcast $t_r^{[k]'}$ to all data holders.
  end
  Data Holders do:
    for $j = 1 \rightarrow i$*, Data holder* j **do**
      Recover the real scores using p:
      $t_r^{[k]} = \frac{t_r^{[k]'}}{p}$
    end
  end
End

3 Experiments

As proof of concept, we applied FedPCA and FedMPCA to two industrial case studies from semiconductor manufacturing. The corresponding datasets SECOM[1] [16] and ST-AWFD[2] [17] have been published previously and are in the public domain.

3.1 General Settings

For both case studies, we first divided the data ($\boldsymbol{X}$) variable-wise into two subsets ($\boldsymbol{X}_1$ and $\boldsymbol{X}_2$) corresponding to different process steps and assigned these to two (hypothetical) data holders. Subsequently, each subset was further split into a training, a validation, and a test set, i.e. $\boldsymbol{X}_1 = \{\boldsymbol{X}_1^{train}, \boldsymbol{X}_1^{val}, \boldsymbol{X}_1^{test}\}$ and $\boldsymbol{X}_2 = \{\boldsymbol{X}_2^{train}, \boldsymbol{X}_2^{val}, \boldsymbol{X}_2^{test}\}$. The partition was done in a way that the training set only contained NOC batches and the validation and test sets consisted of both NOC and faulty batches. The training set was used for training the model. The validation set was used to set control limits for Hotelling's T^2 and Q-statistic. By means of a grid search, we selected the lowest thresholds with the highest F1 score on the validation set as the control limits. The test set was utilized to evaluate the model performance. Four models were built to simulate three common real-world situations:

- Situation 1: One company has access to all data $\boldsymbol{X}$ and can use that data to build a fault detection model based on PCA (MPCA).
- Situation 2: Each company only has access to its data and can use that data to build a local fault detection model to detect faults that occurred in their production line. In this case, Company 1 owns the PCA1 (MPCA1), and Company 2 owns the PCA2 (MPCA2) model. If a fault is detected by one of the models, it is considered to be detected.
- Situation 3: Each company only has access to its data; however, the two companies cooperate to build a federated fault detection model based on FedPCA (FedMPCA).

Table 1 shows the training and test set used for each model. The number of principal components was chosen such that the cumulative sum of explained variance was $\geq 90\%$.

We evaluated the models in terms of the effectiveness to detect faulty batches and also fault diagnosis on the test set. A batch was considered faulty when either the T^2 or the Q-statistic exceeded the predefined control limits. The effectiveness was evaluated by the F1 score, and the fault diagnosis was assessed based on Hotelling's T^2- and Q-contribution plots.

[1] https://archive.ics.uci.edu/ml/datasets/SECOM.

[2] https://github.com/STMicroelectronics/ST AWFD.

Table 1 Training set and test set for each model

Model	Training data	Validation data	Test data
PCA1 (MPCA1)	X_1^{train}	X_1^{val}	X_1^{test}
PCA2 (MPCA2)	X_2^{train}	X_2^{val}	X_2^{test}
PCA (MPCA)	$\{X_1^{train}, X_2^{train}\}$	$\{X_1^{val}, X_2^{val}\}$	$\{X_1^{test}, X_2^{test}\}$
FedPCA (FedMPCA)	X_1^{train}, X_2^{train}	X_1^{val}, X_2^{val}	X_1^{test}, X_2^{test}

Table 2 List of variables that belong to each data holder

Dataset	No. Variables	Variable name
X_1	21	$S_{15}, S_{27}, S_{33}, S_{36}, S_{48}, S_{60}, S_{62}, S_{64}, S_{118}, S_{122}, S_{124}, S_{125}, S_{131}, S_{134}, S_{145}, S_{153}, S_{184}, S_{201}, S_{206}, S_{288}, S_{342}$
X_2	17	$S_{421}, S_{426}, S_{427}, S_{430}, S_{435}, S_{454}, S_{461}, S_{470}, S_{478}, S_{492}, S_{511}, S_{520}, S_{525}, S_{560}, S_{569}, S_{572}, S_{574}$

Table 3 Summary of training and test set

Dataset	No. NOC samples	No. faulty samples	No. features
X_1^{train}	488	0	21
X_1^{test}	74	48	21
X_2^{train}	488	0	17
X_2^{test}	74	48	17

3.2 *Case Study 1: SECOM Dataset*

Data description. SECOM is a static dataset consisting of 1567 observations, each with 590 variables (S_1 to S_{590}) and one label for the quality test (−1 means the observation is normal, and 1 indicates the observation is faulty). As with any real-world dataset, SECOM contains missing values and irrelevant variables. As the focus of this case study is fault detection and diagnosis, we considered the 38 variables recommended by [15] and dropped all instances that contain missing values. According to [15], the selected variables can be divided into five workstations based on the property of the semiconductor manufacturing monitoring process. In this experiment, we assumed that X_1 and X_2 include all parameters of the first three and the last two workstations, respectively. Tables 2 and 3 show a summary of each dataset.

In this experiment, the optional validation set was not used and the control limits for T^2 and Q-statistic set according to Eqs. 6 and 7, respectively.

Table 4 Model performance on SECOM data

Model	TP	TN	FP	FN	F1 score
PCA	19	55	19	29	0.44
FedPCA	19	55	19	29	0.44
PCA1 + PCA2	14	53	21	34	0.33

Evaluation. The performance of all evaluated models is shown in Table 4. PCA and FedPCA show the same performance. This is expected because, in [11], the authors prove that FedSVD is a lossless method that produces the same results as standard SVD. Notably, FedPCA outperforms PCA1, PCA2, and their combination by achieving a higher F1 score which underpins the benefit of integrated versus local process models.

To evaluate the capability of our approach to diagnose faults, we used the trained models to generate contribution plots for Hotelling's T^2 and Q-statistic. Figures 4, 5, and 6 show contribution plots of some selected faults. In each of these figures, there are two plots. The upper plot shows the contribution of each input variable calculated by PCA1 and PCA2 (i.e. the local models). The lower one shows the variable contributions calculated by FedPCA. In the middle of each plot, there is a vertical dashed line representing the (hypothetical) company border. Note that each company can only reconstruct the contributions corresponding to the variables that they own using their private data (x_i), private loadings matrix ($\boldsymbol{V}_i$), and the shared matrix $\boldsymbol{\Sigma}$.

Figures 4 and 5 show two examples of faulty products that were detected by both FedPCA and the combination PCA1 + PCA2. In Fig. 4, it can be seen that for Sample 1, the two plots are quite similar and variables of Company 1 show a much higher contribution to the fault than those of Company 2 indicating that the fault is caused predominantly by the latter. Variables #3 and #4 are reasonable candidates for further root cause analysis. In contrast, in Fig. 5 FedPCA and the local models disagree in terms of the variables that contribute to the fault. Whereas the local models suggest a significant contribution from both parties (with variables from company 2 in fact showing higher overall contributions), the FedPCA model indicates that the fault is mostly associated with company 1. While Variable #14 shows the highest impact according to PCA1 and FedPCA, the most contributed variable according to PCA2 is Variable #32.

Figure 6 demonstrates an example where FedPCA detected a faulty sample that passed both PCA1 and PCA2. Even though the Q-statistic calculated by PCA1 is high, it doesn't surpass the control limit. For PCA2, it is clear that the Q-statistic is small. This example represents cases where the problem is caused by not only the process parameters of one data holder but by a combination/interaction of process parameters across the (hypothetical) company border. An advantage of FedPCA, in this case, is that it can make use of all available data to increase performance, and at the same time, the root cause (sensitive information) is known by the corresponding data holder only, and hidden to the other parties. While the second company can reliably claim the problem did not come from their process production and request the first company to do a checkup. It won't know the exact contribution of input variables of the first company.

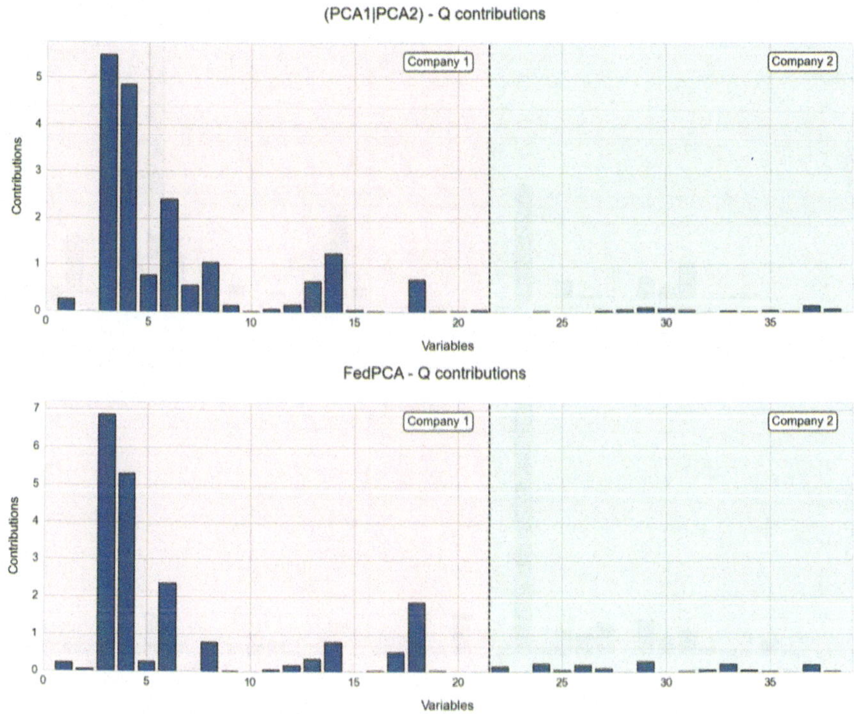

Fig. 4 *Q* contribution plots generated for sample 1

On the other hand, the first company can use the contribution plots together with their know-how to optimize their machine settings to reduce out-of-specification events later in the value chain.

3.3 Case study 2: ST-AWFD Dataset

Data description. ST-AWFD is a batch dataset that contains a total of 1156 batches with 20 variables and an average of 100 samples per batch. Each batch is labeled as normal or faulty through a temporal reference window. The production process is divided into two steps called Step 1 and Step 2. Depending on the batch, the length of each step might differ.

In order to apply batch-wise MPCA and FedMPCA, all the batches must have the same length. Therefore, in this experiment, we have only selected batches with lengths of 110 consisting of 65 observations for Step 1 and 45 observations for Step 2. After the data cleaning phase, there are 966 batches left which include 648 NOC batches and 318 faulty batches. A summary of the data partition is shown in Table 5.

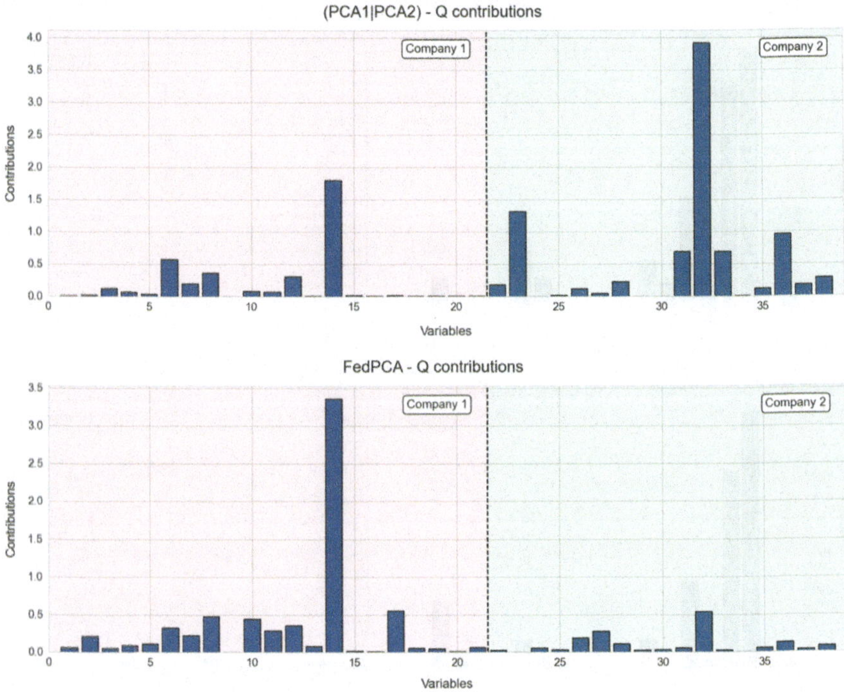

Fig. 5 Q contribution plots generated for sample 2

Table 5 A summary of training, validation, and test set used in the experiment

Dataset	No. NOC samples	No. faulty samples	No. features	No. time intervals
X_1^{train}	482	0	20	65
X_1^{val}	83	159	20	65
X_1^{test}	83	159	20	65
X_2^{train}	482	0	20	45
X_2^{val}	83	159	20	45
X_2^{test}	83	159	20	45

The control limit for the T^2 statistic was calculated by Eq. 6. For the Q-statistic, initially, Eq. 7 was used for determining the confidence limit. However, we found that the returned threshold was too conservative and led to a high number of false positives for all models. Therefore, the control limit for the Q-statistic was obtained by cross-validation using the training and validation set (described in 3.1) instead.

Evaluation The performance of all evaluated models is shown in Table 6. Similar to case study 1, MPCA and FedMPCA returned the same performance and outperformed the combination of MPCA1 and MPCA2. In this experiment, the contribution plots

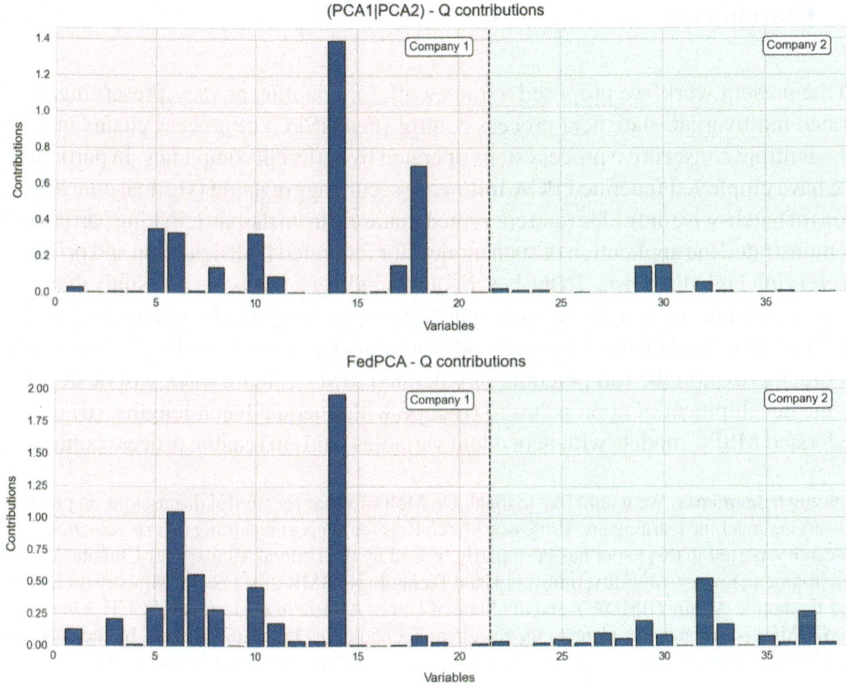

Fig. 6 Q contribution plots generated for sample 3

Table 6 Model performance on ST-AWFD data

Model	TP	TN	FP	FN	F1 score
MPCA	159	83	0	0	1
FedMPCA	159	83	0	0	1
MPCA1 + MPCA2	159	74	9	0	0.97

were also generated for faulty batches. However, because the number of variables is large, it is difficult to judge the difference between the plots without further process knowledge (results not shown).

Altogether, our results on the two case studies underpin the benefits of federated, PCA-based process modeling in terms of better fault detection performance and more informative fault diagnosis that takes into account the interactions between process parameters across (hypothetical) company borders.

4 Conclusion

In the present work, we proposed a framework for enabling privacy-preserving, federated multivariate statistical process control (FedMSPC) of process chains involving multiple consecutive process steps operated by different companies. In particular, we have employed federated PCA following secure aggregation (vertical concatenation) of batch-wise unfolded (and encrypted) datasets from the participating parties and demonstrated the application of such models for federated fault detection and privacy-preserving fault diagnosis. To the best of our knowledge, this is the first study that proposes a solution to this problem. Importantly, our approach provides new incentives for, and underpins the benefits of, closer collaboration of stakeholders along value chains. In order to exploit the full potential of federated MSPC, future work will be devoted to the development of approaches to (i) cope with unequal batch lengths, (ii) derive federated MSPC models with dependent variables, and (iii) enable process control.

Acknowledgements We would like to thank Dr. Mohit Kumar for fruitful discussions on privacy-preserving machine learning and Professor Marco Reis for support regarding dataset selection. The research reported in this paper has been partly funded by the Federal Ministry for Climate Action, Environment, Energy, Mobility, Innovation and Technology (BMK), the Federal Ministry for Digital and Economic Affairs (BMDW), and the State of Upper Austria in the frame of SCCH, a center in the COMET—Competence Centers for Excellent Technologies Program managed by the Austrian Research Promotion Agency FFG and the FFG project circPlast-mr (Grant No. 889843).

References

1. Pagoropoulos A, Pigosso DC, McAloone TC (2017) The emergent role of digital technologies in the circular economy: a review. Proc CIRP 64:19–24
2. Pennekamp J, Henze M, Schmidt S, Niemietz P, Fey M, Trauth D, Bergs T, Brecher C, Wehrle K (2019) Dataflow challenges in an internet of production: a security and privacy perspective. In: Proceedings of the ACM workshop on cyber-physical systems security and privacy, pp. 27–38
3. Voigt P, Bussche A (2017) The EU general data protection regulation (GDPR): a practical guide. https://doi.org/10.1007/978-3-319-57959-7
4. Winning M, Calzadilla A, Bleischwitz R, Nechifor V (2017) Towards a circular economy: insights based on the development of the global ENGAGE-materials model and evidence for the iron and steel industry. Int Econ Econ Policy 14. https://doi.org/10.1007/s10368-017-0385-3
5. Molina-Sánchez E, Leyva-Díaz JC, Cortés-García FJ, Molina-Moreno V (2018) Proposal of sustainability indicators for the waste management from the paper industry within the circular economy model. Water 10:1014. https://doi.org/10.3390/w10081014
6. Konečný J, McMahan HB, Ramage D, Richtárik P (2016) Federated optimization: distributed machine learning for on-device intelligence. arXiv:1610.02527
7. Li L, Yuxi F, Mike T, Kuo-Yi L (2020) A review of applications in federated learning. Comput Ind Eng 149:106854. ISSN 0360-8352. https://doi.org/10.1016/j.cie.2020.106854
8. Qinbin L, Bingsheng H, Dawn S (2021) Proceedings of the IEEE/CVF conference on computer vision and pattern recognition (CVPR), pp 10713–10722

9. Hartebrodt A, Nasirigerdeh R, Blumenthal DB, Röttger R (2021) Federated principal component analysis for genome-wide association studies. In: 2021 IEEE international conference on data mining (ICDM). IEEE, pp 1090–1095
10. Grammenos A, Mendoza Smith R, Crowcroft J, Mascolo C (2020) Federated principal component analysis. Adv Neural Inf Process Syst 33:6453–6464
11. Chai D, Wang L, Fu L, Zhang J, Chen K, Yang Q (2021) Federated singular vector decomposition. arXiv:2105.08925
12. Yang Q, Liu Y, Chen T, Tong Y (2019) Federated machine learning: concept and applications. ACM Trans Intell Syst Technol (TIST) 10(2):1–19
13. Nomikos P, MacGregor JF (1994) Monitoring batch processes using multiway principal component analysis. AIChE J 40(8):1361–1375
14. Nomikos P, MacGregor JF (1995) Multivariate SPC charts for monitoring batch processes. Technometrics 37(1):41–59
15. Arif F, Suryana N, Hussin B (2013) Cascade quality prediction method using multiple PCA+ID3 for multi-stage manufacturing system. IERI Proc. 4:201–207. https://doi.org/10.1016/j.ieri.2013.11.029
16. McCann M, Johnston A (2008) SECOM. UCI Mach Learn Repository
17. Furnari G, Vattiato F, Allegra D, Milotta FLM, Orofino A, Rizzo R, De Palo RA, Stanco F (2021) An ensembled anomaly detector for wafer fault detection. Sensors 21:5465. https://doi.org/10.3390/s21165465
18. Ji Z, Lipton ZC, Elkan C (2014) Differential privacy and machine learning: a survey and review. arxiv:1412.7584v1
19. Bonawitz K, Ivanov V, Kreuter B, Marcedone A, McMahan HB, Patel S, Ramage D, Segal A, Seth K (2016) Practical secure aggregation for federated learning on user-held data. arxiv:1611.04482v1

LPV Modeling for Control Scheme Design of a Compound Helicopter

Bowen Nie, Zhiyin Huang, Long He, Liangquan Wang, and Olivier Sename

Abstract The compound helicopter is able to reach speeds that significantly surpass the conventional counterpart. However, the compounding of lift and thrust always results in more complicated aerodynamic and control issues than a conventional helicopter. Therefore, it is important to model and evaluate the flight dynamics in the early design phase of a compound configuration. The aim of this paper is to develop a Linear-Parameter-Varying (LPV) model of a compound helicopter and investigate the trim, stability and control characteristics. A series of discrete linear state-space models and trim data are obtained from the nonlinear mathematic model, and then interpolated for construction of a LPV model with respect to two scheduling parameters. Lastly, the LPV model is augmented with control scheme to perform flight simulation covering the speed envelope.

Keywords Compound helicopter · LPV modeling · Control scheme · Flight simulation

1 Introduction

The maximum flight speed of conventional helicopter is restricted by adverse aerodynamic effects of stall on the retreating blades and compressibility on the advancing blades of the main rotor [1]. Compounding is a promising solution to increase

B. Nie · Z. Huang (✉) · L. He · L. Wang
Key Laboratory of Rotor Aerodynamics, China Aerodynamics Research and Development Center, Mianyang 621000, China
e-mail: huangzhiyin@cardc.cn

B. Nie
e-mail: niebowen@cardc.cn

O. Sename
GIPSA-Lab, Université Grenoble Alpes, CNRS, Grenoble INP, 3800 Grenoble, France

G. L. Conte and O. Sename (eds.), *Proceedings of the 11th International Conference on Mechatronics and Control Engineering*, Lecture Notes in Mechanical Engineering,
https://doi.org/10.1007/978-981-99-6523-6_4

the maximum flight speed of the helicopter. In recent years, helicopter manufactures, such as Sikorsky and Airbus Helicopters (AH), are exploring and testing the compounding prototypes for future civil and military applications [2].

In practice, both lift and thrust compounding are required to increase the maximum speed of the helicopter. Take the AH X3 as example, the lift compounding is realized with wings to offload the main rotor at high speed, the thrust compounding is equipped with propellers to replace the tail-rotor at low speed and provide propulsive force at high speed. Consequently, the compound helicopter encounters inherent design and modeling challenges, in terms of complicated rotor dynamics, aerodynamic interactions, and redundant controls. To improve the design of compounding configuration at the initial design phase, it is essential to assess the stability, controllability and performance with a full flight-envelope high-fidelity flight dynamic model. Recent studies have performed various approaches to model the flight dynamics of compound helicopter. The practical modeling techniques are composed of low-fidelity models [3, 4] with limited components or neglected mechanics, comprehensive models [5] with multidisciplinary high-fidelity tool chain, and linearized models around the trimmed condition of steady flight [6]. However, the models mentioned in the foregoing sentence are either low-fidelity or sophisticated.

The approach of LPV modeling is adequate to trade off the fidelity and complexity of mathematical model for compound helicopter. In this technique, a set of linear state-space models are obtained by trimming and linearizing the nonlinear high-fidelity model at discrete flight conditions, and then meshed together to from a continuous and time-varying mathematical model covering the entire flight envelope. Application of the LPV framework to fixed wing aircrafts [7, 8], helicopters [9, 10] and tiltrotor aircrafts [11] can be found in literature. However, a rear number of studies have been carried out for compound helicopters. The aim of this paper is to develop a LPV model for compound helicopter representative of AH X3 with two varying parameters: velocity V and rotor speed Ω. The resulting LPV model can be utilized not only for the purpose of full-envelope simulation but also for the Stability and Control Augmentation System (SCAS) design.

The paper is organized as follow. First, a brief theoretical background of LPV and quasi-Linear-Parameter-Varying (qLPV) is introduced. Second, a nonlinear mathematical dynamic model of the studied compound helicopter is presented. Next, a LPV model is constructed by scheduling the state-space models and corresponding trim values for a set of grid points. Lastly, a SCAS is designed and applied to the g LPV model for closed-loop simulation covering the speed envelope.

2 Theoretical Background

LPV systems are linear state-space models that depend on a time varying parameter vector $\rho(t)$. Namely, a collection of linear state-space models and the corresponding trim data, obtained at a set of discrete equilibrium points, can be combined into lookup tables and interpolated as function of the scheduling parameters [12]. As a result, a LPV system is defined as following:

$$\dot{X}(t) = A(\rho(t))(X(t) - X_0(\rho(t))) + B(\rho(t))(U(t) - U_0(\rho(t))) \tag{1}$$

$$Y(t) = C(\rho(t))(X(t) - X_0(\rho(t))) + D(\rho(t))(U(t) - U_0(\rho(t))) \tag{2}$$

where X, U and Y are the state, input and output vectors, X_0, U_0 and Y_0 are the operating point state, input and output vectors, respectively.

A qLPV system is a particular case of LPV system, when a subset of scheduling parameter is also state of system. Namely, the state vector $X = [Z, W]^T$ is composed of scheduling states $Z \subset \rho(t)$ and non-scheduling states $W \not\subset \rho(t)$. Consequently, the LPV system of Eq. (1) can be rewritten as:

$$\begin{bmatrix} \dot{Z} \\ \dot{W} \end{bmatrix} = \begin{bmatrix} A_{11}(\rho) & A_{12}(\rho) \\ A_{21}(\rho) & A_{22}(\rho) \end{bmatrix} \begin{bmatrix} Z - Z_0(\rho) \\ W - W_0(\rho) \end{bmatrix} + \begin{bmatrix} B_1(\rho) \\ B_2(\rho) \end{bmatrix} [U - U_0(\rho)] \tag{3}$$

Since Z is used in the scheduling parameter function $\rho(t)$ and also state of the system, $Z - Z_0(\rho) = 0$ is always true. Therefore, Eq. (3) can be simplified as following:

$$\begin{bmatrix} \dot{Z} \\ \dot{W} \end{bmatrix} = \begin{bmatrix} A_{11}(\rho) & A_{12}(\rho) \\ A_{21}(\rho) & A_{22}(\rho) \end{bmatrix} \begin{bmatrix} 0 \\ W - W_0(\rho) \end{bmatrix} + \begin{bmatrix} B_1(\rho) \\ B_2(\rho) \end{bmatrix} [U - U_0(\rho)] \tag{4}$$

However, the contributions of scheduling states to $\dot{Z}$ can be reserved implicitly through the variation of trim states and inputs as described in reference [13].

3 Compound Helicopter Mechanics

3.1 *Hybrid Compounding Configuration*

The studied helicopter employs hybrid compounding with both lift and thrust. Lift is generated by the main rotor and a jointed box wing, and thrust is provided simultaneously by the main rotor and a pair of lateral propellers mounted on the wing tips. Since auxiliary lift and thrust can be obtained at high speed, the hybrid compounding configuration is expected to reach a potential flight speed of V = 68 m/s.

Regarding the control surfaces of the studied compound helicopter, a mean collective pitch P_a of the propellers is responsible for the thrust, while a differential collective pitch P_d of the propellers is required to counteracts the anti-torque of the main rotor and addresses the yaw damping. In addition, an H-stabilizer helps to provide the horizontal, vertical static stability and rudder δ_r. Besides, there are three control surfaces of the main rotor including the collective θ_0, , longitudinal cyclic pitch (B_1) and lateral cyclic pitch (A_1).

3.2 Nonlinear Dynamic Equations

To develop the nonlinear flight dynamic model, a summarization of the forces F^B and moments M^B with respect to the center of gravity in the body frame is required.

$$\begin{pmatrix} mI_3 & 0 \\ 0 & J \end{pmatrix}\begin{pmatrix} \dot{\vec{v}}^B \\ \dot{\vec{\omega}}^B \end{pmatrix} + \begin{pmatrix} \vec{\omega}^B \times m\vec{v}^B \\ \vec{\omega}^B \times J\vec{\omega}^B \end{pmatrix} = \begin{pmatrix} F^B \\ M^B \end{pmatrix} \quad (5)$$

where m is the mass, J is the inertia tensor, $\vec{v}^B = [u, v, w]^T$ and $\vec{\omega}^B = [p, q, r]^T$ are the translational and angular speed in the body-fixed frame, respectively.

As the compound helicopter is composed of several subsystems, the forces and moments can be decomposed as following:

$$\begin{cases} F^{\mathrm{B}} = F_g + F_r + F_p + F_f + F_w + F_s + F_t \\ M^{\mathrm{B}} = M_r + M_p + M_f + M_w + M_s + M_t \end{cases} \quad (6)$$

In Eq. (6), the subscripts of g, r, p, f, w, s and t denote the gravity, rotor, propeller, fuselage, wing, stabilizer and vertical tail, respectively.

To account for the nonlinear aerodynamics and rotor periodicity, an 'individual blade model' is developed for the main rotor and propellers using the approach described in reference [14]. Aerodynamic loads of the fuselage, wing and fins are obtained by a series of lookup tables and interpolations with the experimental data. Furthermore, dynamic inflow model and a rotor-speed governor model are also integrated to the nonlinear compound helicopter dynamics as described in reference [15, 16].

From a control point of view, the entire compound helicopter dynamics take the nonlinear form as:

$$\dot{\vec{x}} = f\left(\vec{x}, \vec{u}\right) \quad (7)$$

where $\vec{x} = [\Delta u, \Delta v, \Delta w, \Delta p, \Delta q, \Delta r, \Delta\phi, \Delta\theta, \Delta\psi]^T$ is the perturbation state vector composed of six rigid body speeds and three Euler angles (ϕ, θ, ψ), $\vec{u} = [\Delta A_1, \Delta B_1, \Delta P_d, \Delta\theta_0, \Delta P_a, \Delta\delta_r]$ is the perturbation control vector.

4 LPV Modelling Step

4.1 State-Space Point Models for the Compound Helicopter

In this paper, a LPV model is developed for a compound helicopter using two scheduling parameters: velocity V and rotor speed Ω. Velocity is selected as a scheduling parameter to capture the changing dynamics, such that the entire speed envelope can be covered. Rotor speed is chosen as an additional scheduling parameter to account for the dynamics induced by the rotor speed, which should be slow down to offload the main rotor as shown in Fig. 1.

The final choice of the scheduling parameters is given in Table 1. According to the two-dimensional scheduling network, there are 16 grid points. A collection of linear state-space models and the corresponding trim data is then generated for all the grid points. The resulting state-space models take the form of:

$$\sum_i = \begin{cases} \dot{\vec{x}} = A_i \vec{x} + B_i \vec{u} \\ \vec{y} = C_i \vec{x} + D_i \vec{u} \end{cases} \quad i = 1, 2, \ldots 16 \tag{8}$$

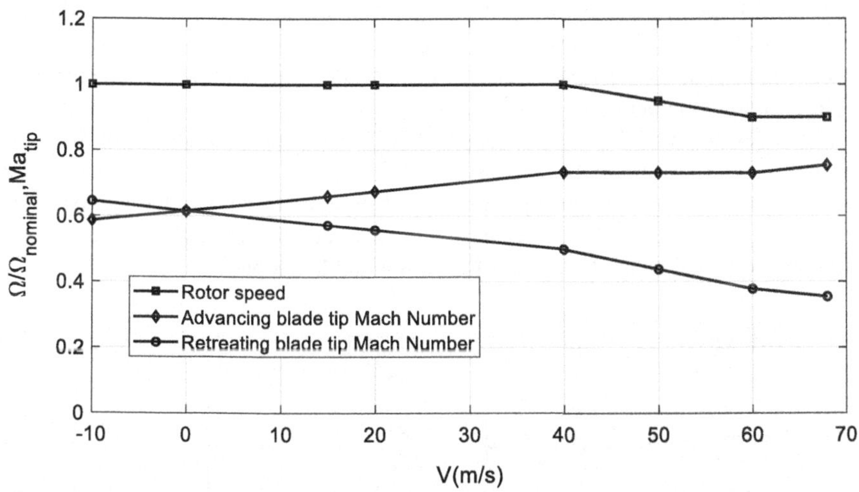

Fig. 1 Variation of rotor speed for the compound helicopter

Table 1 Scheduling parameters

Parameter	Values	Unit
Velocity V	[−10, 0, 15, 20, 40, 50, 60, 68]	m/s
Rotor speed Ω	[900, 1000]	rpm

where $\vec{x}$ and $\vec{u}$ are the same state and input vectors as Eq. (7), and $\vec{y} = \vec{x}$ is the output vector for sensor feedback.

4.2 LPV Model Structure for the Compound Helicopter

A block diagram schematic of the LPV model is shown in Fig. 2. Note that the aforementioned state-space point models are scheduled with $\rho = [V, \Omega]$, and then model meshing is implemented through lookup tables and interpolations. First, the lookup tables of trim control inputs, trim states and stability & control derivatives are generated as function of the scheduling parameters. Then, the interpolated trim control and trim states are subtracted from the current values to obtain the perturbations. Lastly, the control and state perturbations are multiplied with the interpolated control and stability matrices to calculate the state accelerations, which will be further integrated to obtain the current states.

Figure 3 presents an example for comparison of the trim values between the LPV model and the linear point models in terms of the pitch angle θ, rotor collective θ_0 and differential propeller collective P_d, which are captured off the grid nodes listed in Table 1. It is noted that the gain-scheduling LPV trim values show a good match with the linear point models across the flight speed envelope. As the flight speed increases, the wing offloads the main rotor and the vertical tail offloads the antitorque gradually. It is reasonable that the required trim values of θ, θ_0 and P_d decrease at high flight speed. This validates the LPV model, which captures the nonlinear characteristics of the compound helicopter.

Moreover, the modal characteristics of LPV model are calculated over the fight speed range -10 to 68 m/s at increments of 1 m/s, and compared with that of the linear point models off the grid points. It is observed that modal characteristics of the two models agree well with each other. This validates the LPV model, which captures the modal characteristics of the compound helicopter.

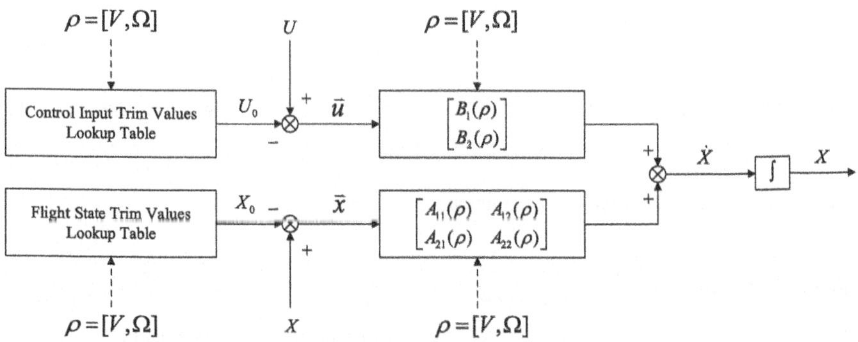

Fig. 2 LPV model structure for the compound helicopter

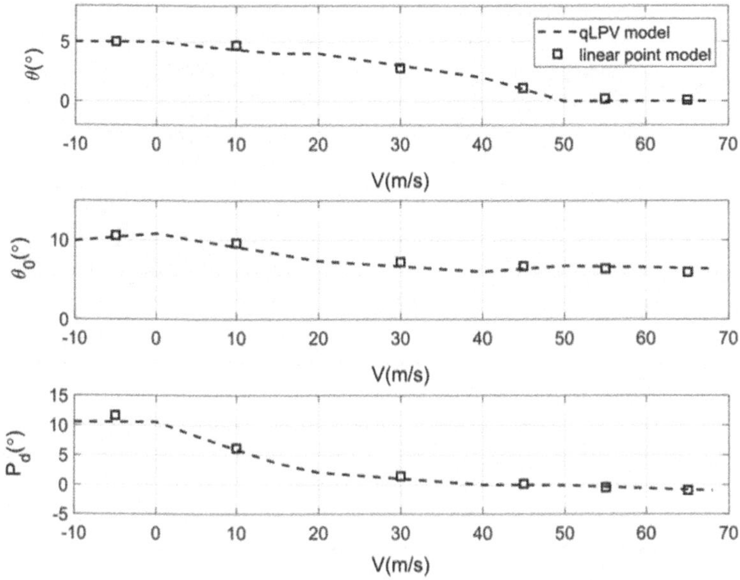

Fig. 3 Trim results of the nonlinear and LPV model

Figure 4a shows the longitudinal modes of the compound helicopter, including the phugoid, short period, pitch subsidence and heave subsidence. As can be seen, the pitch and heave subsidence modes combine to form short period mode as the flight speed increases. To be mentioned is that the phugoid mode is unstable at low speed while the damping of short period mode decreases at high flight speed. The lateral and directional modes are demonstrated in Fig. 4b. Similar to a conventional helicopter, the dutch roll mode is unstable at hover. As the speed increases, the dutch roll mode becomes stable. However, the damping of dutch roll mode decreases at high speed.

To conclude, it can be inferred that a SCAS is required to stabilize and control the compound helicopter over the speed envelope.

5 Simulation of LPV Model

5.1 Stability and Control Augmentation Scheme

To improve the stability and performance of the compound helicopter, a SCAS is designed based on the linear point models and then applied to augment the LPV model. The proposed SCAS scheme is presented in Fig. 5, which is composed of the longitudinal and lateral & directional channels.

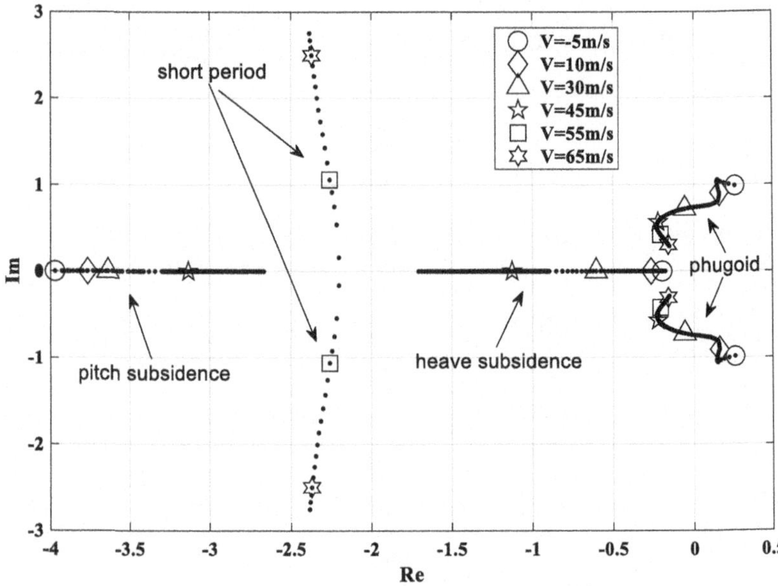

(a) Longitudinal modal characteristics of the compound helicopter.

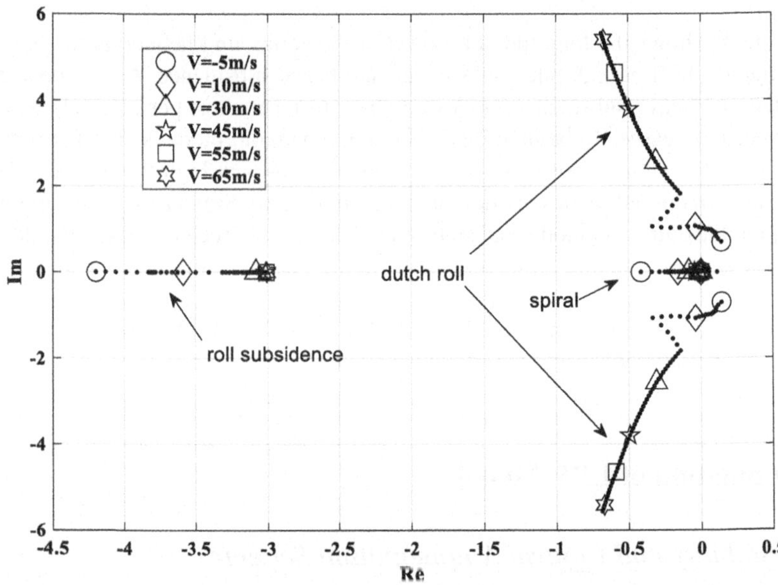

(b) Lateral and directional modal characteristics of the compound helicopter.

Fig. 4 Modal characteristics of the compound helicopter

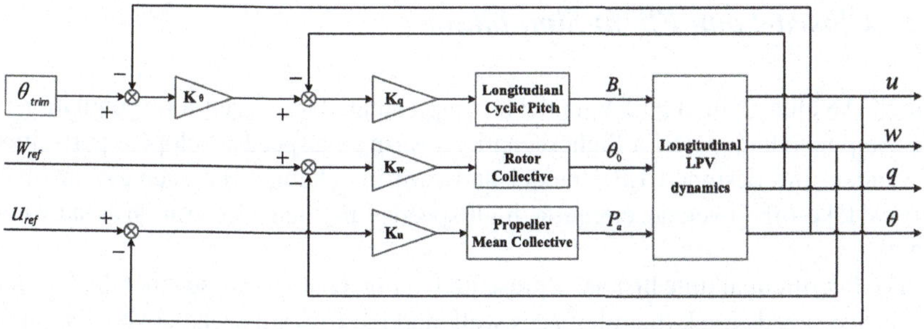

(a) Longitudinal control scheme for the compound helicopter.

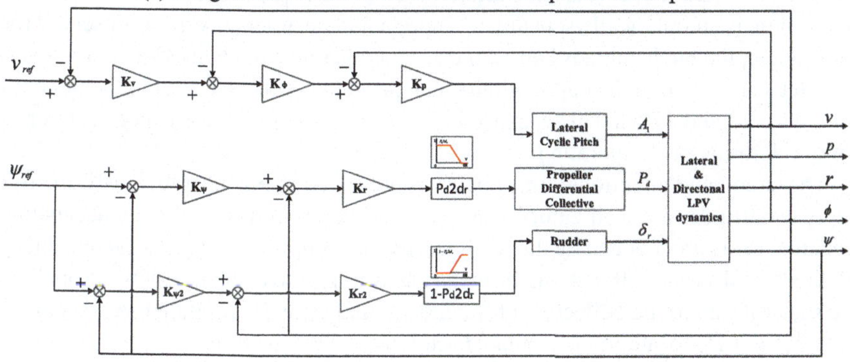

(b) Lateral & directional control scheme for the compound helicopter.

Fig. 5 Stability and control scheme for the compound helicopter

As demonstrated in Fig. 5a, the pitch rate feedback is introduced to improve the pitch damping, while the feedback loops of pitch angle, forward speed and heave speed are employed to track the references, respectively. Since the forward speed can be driven through the mean collective of the propellers, the forward speed is decoupled with the pitch attitude, and the trim values of pitch angel are selected as the reference. Furthermore, the feedback gains and θ_{trim} are scheduled to the flight speed.

The lateral & directional control scheme is shown in Fig. 5b. The roll rate, roll angle and lateral speed are cascaded to track the reference of lateral speed, while the yaw damping and yaw angle tracking loops are allocated to the differential propeller collective and the rudder, simultaneously. Actually, the commands of P_d and δ_r are scaled with two nondimensional factors, which are scheduled with the flight speed in the section of 0–1.

5.2 Closed-Loop Flight Simulation

The SCAS along with the LPV model are implemented for closed-loop simulation of the compound helicopter. A flight scenario covering the speed envelope is performed to validate the designed LPV model and control scheme. The scenario involves vertical take-off, hover, acceleration, high-speed cruise and deceleration as shown in Fig. 6.

The longitudinal time history is presented in Fig. 6a. One can see that the forward and heave speeds track the references well, and the forward speed covers the range from 0 to 68 m/s. Following the variation of forward speed, the reference value of pitch angle is automatically scheduled. Though notable tracking error is observed for pitch angle, the pitch angle is kept in the acceptable range of about 3°–6°. Moreover, clip steps are found in the curve of the mean propeller collective at the time of about 50, 100, 120 and 170 s. This is caused by the variation of forward speed reference for accelerating or decelerating.

The lateral and directional time history is presented in Fig. 6b. Obviously, the roll and yaw attitudes are well damped and always kept around zero, though acceptable tracking errors are induced by the nonlinearity and coupling. To verify the feasibility of directional surface allocation, the mean and differential collective of the propellers are transformed to the collective of the left P_L and right P_R propeller. As is shown, both P_L and P_R locate in the available section of $-15°$ to $+40°$.

In summary, the time history revels that the LPV model is adequate to capture the dynamics of the compound helicopter and the designed SCAS is effective to stabilize and control the compound helicopter throughout the speed envelope.

6 Conclusion

This paper presented the development of LPV model for a compound helicopter to trade off the fidelity and complexity. The state-space models and the corresponding trim values are scheduled with respect to the varying parameters of velocity and rotor speed. The implemented LPV model agrees well with the nonlinear dynamic model in terms of trim value and modal characteristics both on and off the scheduling grid points.

A SCAS is designed and applied to the LPV model for closed-loop flight simulation. A scenario involving vertical take-off, hover, acceleration, cruise and deceleration is performed to validate the dynamics and control scheme. The time history of flight simulation across the flight speed envelope revels that the LPV model is adequate to capture the nonlinear dynamic characteristics of the compound helicopter, and the designed SCSA is effective to stabilize and control the compound helicopter.

In the future, research focus will be placed on implementing the LPV model and SCAS in experimental setup for further validation and improvement.

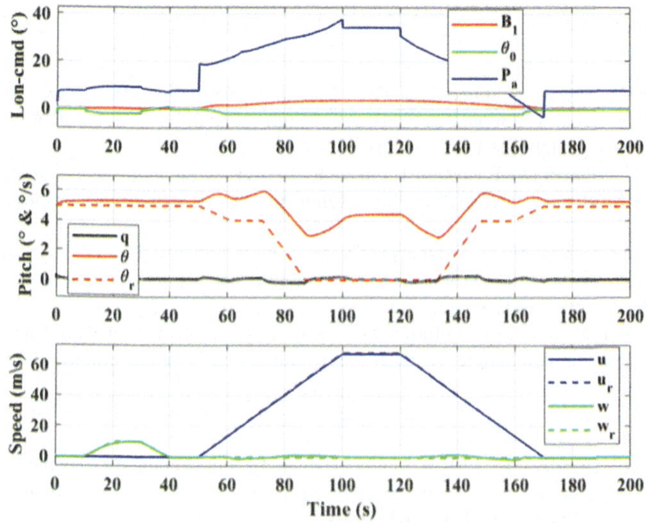

(a) Longitudinal time history.

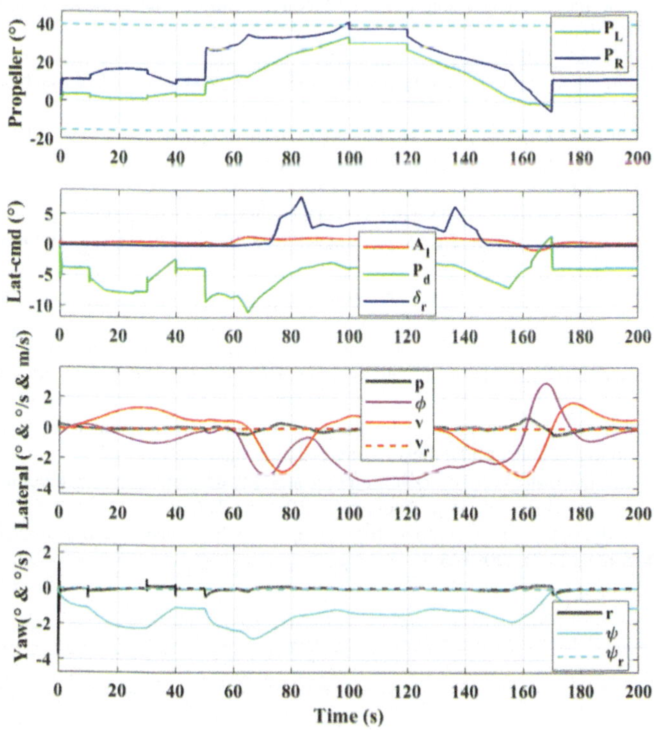

(b) Lateral & directional time history.

Fig. 6 Time history of the flight simulation over the speed envelope

References

1. Filippone A (2006) Flight performance of fixed and rotary wing aircraft, CA
2. Hrle C et al (2020) Compound helicopter X in high-speed flight: correlation of simulation and flight test. J Am Helicopter Soc. 66(1)
3. Lienard C, Salah el Din I, Renaud T, Fukari R (2018) RACER high-speed demonstrator: rotor and rotor-head wake interactions with tail unit. In: American helicopter society 74th annual forum
4. Stokkermans T, Voskuijl M, Veldhuis L, Soemarwoto B, Fukari R, Eglin P (2018) Aerodynamic installation effects of lateral rotors on a novel compound helicopter configuration. In: American helicopter society 74th annual forum
5. Thiemeier J et al (2020) Aerodynamics and flight mechanics analysis of Airbus Helicopters' compound helicopter RACER in hover under crosswind conditions. CEAS Aeronaut J 11(1):49–66
6. Panza S et al (2019) QLPV modelling of helicopter dynamics. IFAC-PapersOnLine 52(28):82–87
7. Tobias EL, Tischler MB (2016) A model stitching architecture for continuous full flight-envelope simulation of fixed-wing aircraft and rotorcraft from discrete-point linear models. U.S. ArmyAMRDEC SR RDMR-AF-16-01
8. Berger T, Tischler MB, Cotting MC, Gray WR (2020) Identification of a full-envelope Learjet-25 simulation model using a stitching architecture. J Guide Control Dyn. https://doi.org/10.2514/1.G005094
9. Zivan L, Tischler MB (2010) Development of a full flight envelope helicopter simulation using system identification. J Am Helicopter Soc 55(2):22003. https://doi.org/10.4050/JAHS.55.022003
10. Berger T, Juhasz O, Lopez MJS, Tischler MB, Horn JF (2020) Modeling and control of lift offset coaxial and tiltrotor rotorcraft. CEAS Aeronaut J 11(1):191. https://doi.org/10.1007/s13272-019-00414-0
11. Nabi HN, Visser CCD, Pavel MD et al (2021) Development of a quasi-linear parameter varying model for a tiltrotor aircraft. CEAS Aeronaut J 12:879–894 (2021). https://doi.org/10.1007/s13272-021-00539-1
12. Marcos A, Balas GJ (2007) Development of linear-parameter-varying models for aircraft. J Guidance Control Dyn 27(2):218–228
13. Tischler MB (2012) Aircraft and rotorcraft system identification: engineering methods with flight test examples, 2nd edn. American Institute of Aeronautics and Astronautics Inc., Reston, VA
14. Rutherford S (1997) Simulation techniques for the study of the maneuvering of advanced rotorcraft configurations, Ph.D. thesis. University of Glasgow, UK
15. Ferguson K, Thomson D (2016) Examining the stability derivatives of a compound helicopter. Aeronaut J 121(1235):1–20. https://doi.org/10.1017/aer.2016.101
16. Li et al (2010) A mathematical model for helicopter comprehensive analysis—ScienceDirect. Chin J Aeronaut 23(3):320–326

Piecewise Modeling Algorithms Using Numerical Data

Tadanari Taniguchi and Michio Sugeno

Abstract In this study, a novel piecewise modeling method was devised using numerical data. The piecewise model was represented as a rectangular region divided into a state-space. The vertex values of the rectangular region were determined using the learning algorithm based on a simplified fuzzy inference model because the piecewise model was represented by a fuzzy if-then rule with singleton consequents. The proposed algorithm can be used to determine optimal vertex values and positions of segmented regions and with minimum modeling errors. Three examples were considered to demonstrate the effectiveness of the proposed method using numerical simulations.

Keywords Piecewise nonlinear model · Fuzzy if-then rule · Modeling algorithm

1 Introduction

In the 1950s, Zadeh proposed system identification [12, 13]. Subsequently, prediction error, and subspace identification methods were proposed in 1960s. The maximum likelihood estimation method was used for realizing auto regressive moving average with exogenous (ARMAX) model, which is the basis for the prediction error method [7]. Until 1980s, Ljung summarized prediction error methods [5]. Realization theorems of the state-space model, which is the basis of the subspace method, was proposed by Ho and Kalman [2].

Nonlinear system identification methods [8] have been widely studied. Many system identification methods based on machine learning and data science have been studied as data-driven control system design methods [8]. Piecewise affine

---

T. Taniguchi (✉)
STEM Education Center, Tokai University, Kanagawa 2591292, Japan
e-mail: taniguchi@tokai-u.jp

M. Sugeno
Tokyo Institute of Technology, Tokyo 1528550, Japan

G. L. Conte and O. Sename (eds.), *Proceedings of the 11th International Conference on Mechatronics and Control Engineering*, Lecture Notes in Mechanical Engineering,
https://doi.org/10.1007/978-981-99-6523-6_5

autoregressive exogenous (PWARX) models [1, 6] for hybrid systems have been proposed as system identification methods.

In this study, a piecewise multilinear (PML) model [9] was used as a control system model. The PML system, which is a piecewise nonlinear system, has general approximation capability for nonlinear systems and is fully parametric. PML models are constructed using hypercubes partitioned in the state space and is the second simplest after the piecewise linear model. The adjacent piecewise regions are continuous in PML systems. We derived stabilizing conditions based on feedback linearization for continuous-time and discrete-time systems [10, 11], respectively.

This study entailed the development of a piecewise modeling algorithm using numerical data. The vertex values of the rectangular region can be determined using the learning algorithm based on a simplified fuzzy inference model because the piecewise model is represented by a fuzzy if-then rule with singleton consequents. The proposed algorithm can be used to determine optimal vertex values and positions of segmented regions with minimum modeling errors. Determining the vertex values of rectangular regions and the position of the region segmentations is a nonlinear programming problem. Therefore, it is very difficult to obtain global minimum solution. Because in this method, a piecewise model is used, a model with fewer modeling errors can be constructed by increasing the number of piecewise regions. Finally, three numerical simulations were performed to demonstrate the effectiveness of the proposed method.

2 Canonical Form of the PML Model

2.1 State-Space Expression

An n-dimensional PML system was considered. The state space in n-dimensional case is categorized into

$$\begin{aligned}
x_1 &\in \{d_1(1), \ldots, d_1(r_1), d_1(r_1+1), \ldots, d_1(m_1)\},\\
x_2 &\in \{d_2(1), \ldots, d_2(r_2), d_2(r_2+1), \ldots, d_2(m_2)\},\\
&\vdots\\
x_n &\in \{d_n(1), \ldots, d_n(r_n), d_n(r_n+1), \ldots, d_n(m_n)\},
\end{aligned} \tag{1}$$

where $d_j(i_j)$ is the vertex of x_j, and m_j is the number of vertices of x_j, $j = 1, \ldots, n$. $x \in S$, where S is the bounded region by (1).

The number of piecewise regions is $\prod_{i=1}^{n}(m_i - 1)$. The PML model in the region

$$R_{r_1 \ldots r_n} = [d_1(r_1), d_1(r_1+1)] \times \cdots \times [d_n(r_n), d_n(r_n+1)]$$

is constructed as follows:

$$x^+ = \sum_{i_1=r_1}^{r_1+1} \cdots \sum_{i_n=r_n}^{r_n+1} \omega_1^{i_1}(x_1)\cdots\omega_n^{i_n}(x_n) f(i_1,\ldots,i_n), \tag{2}$$

$$x = \sum_{i_1=r_1}^{r_1+1} \cdots \sum_{i_n=r_n}^{r_n+1} \omega_1^{i_1}(x_1)\cdots\omega_n^{i_n}(x_n) d(i_1,\ldots,i_n),$$

where $x^+ = \left(x_1^+,\ldots,x_n^+\right)^T, x = \left(x_1,\ldots,x_n\right)^T,$

$$f(r_1,\ldots,r_n) = \left(f_1(r_1,\ldots,r_n),\ldots, f_n(r_1,\ldots,r_n)\right)^T,$$
$$d(r_1,\ldots,r_n) = \left(d_1(r_1),\ldots,d_n(r_n)\right)^T.$$

The triangular membership functions are as follows:

$$\omega_1^{r_1}(x_1) = \begin{cases} \dfrac{d_1(r_1+1)-x_1}{d_1(r_1+1)-d_1(r_1)} & d_1(r_1) \le x_1 \le d_1(r_1+1), \\ 0 & \text{otherwise}, \end{cases}$$

$$\omega_1^{r_1+1}(x_1) = \begin{cases} \dfrac{x_1-d_1(r_1)}{d_1(r_1+1)-d_1(r_1)}, & d_1(r_1) \le x_1 \le d_1(r_1+1), \\ 0 & \text{otherwise}, \end{cases}$$

$$\vdots$$

$$\omega_n^{r_n}(x_n) = \begin{cases} \dfrac{d_n(r_n+1)-x_n}{d_n(r_n+1)-d_n(r_n)}, & d_n(r_n) \le x_n \le d_n(r_n+1), \\ 0 & \text{otherwise}, \end{cases}$$

$$\omega_n^{r_n+1}(x_n) = \begin{cases} \dfrac{x_n-d_n(r_n)}{d_n(r_n+1)-d_n(r_n)}, & d_n(r_n) \le x_n \le d_n(r_n+1), \\ 0 & \text{otherwise}. \end{cases}$$

If PML system (2) has continuous-time dynamics, x^+ is considered as $\dot{x}$; if it has discrete-time dynamics, x^+ is considered to be $x(t+1)$.

2.2 *If-Then Rule Expression*

The PML model (2) in $R_{r_1\ldots r_n}$ can be transformed into an if-then rule expression based on a fuzzy reasoning.

$$\begin{aligned}&\text{Rule 1: If } x_1 \text{ is } A_1^{r_1} \text{ and } \cdots \text{ and } x_n \text{ is } A_n^{r_n}\\&\qquad\text{then } x^+ \text{ is } f(r_1, \ldots, r_n)\\&\qquad\vdots\\&\text{Rule } 2^n: \text{If } x_1 \text{ is } A_1^{r_1+1} \text{ and} \cdots \text{ and } x_n \text{ is } A_n^{r_n+1}\\&\qquad\text{then } x^+ \text{is } f(r_1+1, \ldots, r_n+1)\end{aligned} \tag{3}$$

where $A_j^{r_j}$ is the fuzzy set, and the membership function is $\omega_j^{r_j}(x_j)$, $j = 1, \ldots, n$. The degree of the rules is represented as follows:

$$\mu_k(x) = \prod_{j=1}^{n} \omega_j^{i_j}(x_j) = \omega_1^{i_1}(x_1) \cdot \omega_2^{i_2}(x_2) \cdot \cdots \cdot \omega_n^{i_n}(x_n),$$

where $i_1 = r_1, r_1+1, i_2 = r_2, r_2+1, \ldots, i_n = r_n, r_n+1, k = \sum_{j=1}^{n} 2^{j-1}(i_j - r_j + 1)$.

Region $R_{r_1 \ldots r_n}$ has 2^n if-then rules. The PML model has $\prod_{i=1}^{n}(m_i - 1)$ regions, the entire system has $2^n \prod_{i=1}^{n}(m_i - 1)$ rules.

The fuzzy inference system x^+ is obtained as follows:

$$\begin{aligned}x^+ &= \sum_{k=1}^{2^n} \mu_k(x) f^k = \sum_{i_1=r_1}^{r_1+1} \cdots \sum_{i_n=r_n}^{r_n+1} \omega_1^{i_1}(x_1) \cdots \omega_n^{i_n}(x_n) f(i_1, \ldots, i_n),\\ k &= \sum_{i=1}^{n} 2^{i-1}(i_j - r_j + 1),\end{aligned} \tag{4}$$

where $f^k = f(i_1, \ldots, i_n)$. Therefore, the fuzzy inference system (4) is the same as that of the PML system (2).

3 Piecewise Modeling Algorithms

This section details the optimal modeling of the PML system (2). Three variables should be determined for the modeling method:

(1) vertex values $f(i_1, \ldots, i_n)$s of the PML model (2)
(2) the number of piecewise regions: $\prod_{i=1}^{n}(m_i - 1)$
(3) the divided positions $d_j(i_j)$ in the state-space variable x_j,

where $j = 1, \ldots, n$. Because a piecewise model is considered herein, a model with fewer modeling errors can be constructed by increasing the number of piecewise regions. Therefore, two variables namely $f(i_1, \ldots, i_n)$ were optimized, and the positions $d_j(i_j)$ were divided by fixing the number of piecewise regions. However, deriving optimal solutions is difficult because the optimization of two variables is a nonlinear programming problem.

3.1 Algorithm for Finding the Optimal Vertex Values $f(i_1, \ldots, i_n)$

We address the optimal solution of $f(i_1, \ldots, i_n)$ by fixing the number of piecewise regions and divided positions. Using the PML model (2) with nonlinearities is difficult. Therefore, the if-then rule expression (3), which is equivalent to the PML model (2), is used to optimize $f(i_1, \ldots, i_n)$.

$f(i_1, \ldots, i_n)$ can be determined using learning algorithm [4] based on a simplified fuzzy inference model, because the piecewise model is represented by a fuzzy if-then rule with singleton consequents. Algorithm 1 reveals the optimal values $f^k = f(i_1, \ldots, i_n)$ based on the gradient descent method.

Algorithm 1 Optimization of values f^ks

Step 1 Generate training data $(x_*(t), x_*^+(t))$, $t = 1, \ldots, N$.
Step 2 Set the initial data f^k, training rate τ, and number of iterations $i = 1$ to M
Step 3 Apply the training data $x_*(t)$ to the system

$$x^+(t) - \sum_{k=1}^{2^n} \mu_k(x_*(t)) f^k \tag{5}$$

Step 4 Calculate the error $\delta(t)$ between $x^+(t)$ and $x_*^+(t)$

$$\delta(t) = x^{+}(t) - x_*^{+}(t).$$

Step 5 Solve f^k from the equations

$$f^k(t+1) = f^k(t) + \tau \mu_k(x_*(t))\delta(t)$$

using the gradient descent method.
Step 6 Set $t \leftarrow +1$. If $t \leq N$, return to Step 3.
Step 7 Set $i \leftarrow +1$. If $i \leq M$, return to Step 3. Otherwise, then the algorithm is terminated.

The system x^+ in (5) is linear with respect to f^k. Therefore, this algorithm realizes optimal solutions f^ks using the aforementioned iterative calculations. The following equation is used as an evaluation function to measure modeling performance:

$$E = \frac{1}{2N} \sum_{t=1}^{N} (x^+(t) - x_*^+(t))^2, \tag{6}$$

where N denotes the number of training data.

3.2 *Algorithm for Determining the Optimal Solutions of Vertex Values and Divided Positions*

A novel algorithm was proposed to determine the optimal solutions of $f(i_1, \ldots, i_n)$, and the position $d_j(i_j)$ was divided by fixing the number of piecewise regions. This algorithm is based on the spline-function minimization algorithm proposed in [3], which is a fusion of Algorithm 1. Algorithm 2 determines optimal values $f(i_1, \ldots, i_n)$s and positions of region segmentation with minimum modeling errors. Determining the vertex values of rectangular regions and their position of the region segmentations is a nonlinear programming problem. Therefore, this algorithm can determine a local optimal solution. Because in this method, a piecewise model is used a model with fewer modeling errors can be constructed by increasing the number of piecewise regions. In this algorithm, α and β are the training data intervals. The squared errors are represented as follows:

$$E_i^j = \sum_{\substack{x \in S \setminus \{x_j\} \\ d_j(i) \le x_j \le d_j(i+1)}} \frac{1}{2}\left(x^+ - x_*^+\right)^2, \; \bar{E}^j = \sum_{i=1}^{m_j-1} \bar{E}_i^j,$$

$$\bar{E}_i^j = \sum_{x \in S \setminus \{x_j\}, \{\bar{x}_j\}} \frac{1}{2}\left(x^+ - x_*^+\right)^2,$$

where S is the bounded region by (1).

Algorithm 2 Optimization of values f^ks and divided positions $d_j(i_j)$s

Step 1 Set the vertices $x_j \in \{d_j(2), \ldots, d_j(m_j - 1)\}$ in (1).
Step 2 Set $j \leftarrow 1$.
Step 3 Set $i \leftarrow 1$.
Step 4 Calculate the squared errors E_i^j under optimal solution f^k using Algorithm 1
If $E_i^j > E_{i+1}^j$, $d_j'(i+1) = d_j(i+1) - \alpha$.
If $E_i^j < E_{i+1}^j$, $d_j'(i+1) = d_j(i+1) + \beta$.
Calculate the squared error $\bar{E}^j$ under optimal solution f^k using Algorithm 1 in the following case: $\bar{x}_j \in \{d_j(1), \ldots, d_j(i), d_j'(i+1), d_j(i+2) \ldots, d_j(m_j - 1)\}$.
If $\bar{E}^j < \sum_{i=1}^{m_j-1} E_i^j$, then $d_j(i+1) = d_j'(i+1)$.
Step 5 Set $i \leftarrow +1$. If $i = m_1 - 1$ and the vertices d are not updated, proceed to the next step. Otherwise, return to Step 4.
Step 6 Set $j \leftarrow +1$. If $j = n$ and the vertices d are not updated, terminate the algorithm. Otherwise, return to Step 3.

---

4 Simulation Results

In these simulations, we used MATLAB R2022b, which is one of a numerical simulation software on a Windows PC with Intel Xeon W-2225.

4.1 One Dimensional System

A PML model was constructed from numerical data ($N = 61$) using the proposed algorithm. Table 1 lists the numerical data obtained using

$$y = \sin^3 x. \tag{7}$$

The learning parameters $\tau = 1, \alpha = 0.1$, and $\beta = 0.1$ are used as simulation parameters. The upper and lower tables of in Table 2 detail the initial and optimized vertex points, respectively. The initial vertex points, designated as y, were generated by substituting the initial values, x, into (7).

The dotted points in Fig. 1 represent the data in Table 1. The solid and dashed lines indicate PML models using the optimal and initial vertex points in Table 2. The computational time was 7.121×10^{-2} s. The modeling error of the PML model with the optimized vertex points is 0.0033. However, the error when using the initial vertex point is 0.0136. Therefore, the PML model using the proposed algorithm exhibits an excellent modeling performance for the training data.

Table 1 Data of example 1 ($N = 61$)

x	-3	-2.9	-2.8	-2.7	-2.6	...	2.9	3
y	–0.0028	–0.0137	–0.0376	–0.0781	–0.1370	...	0.0137	0.0028

Table 2 Initial and optimized vertex values of example 1

Initial vertex values

x	–3	–2	–1	0	1	2	3
y	–0.0028	–0.7518	–0.5958	0	0.5958	0.7518	0.0028

Optimized vertex values

x	–3	–2	–1.4	–0.5	0.5	1.5	3
y	0.2939	–0.8818	–0.9569	–0.1321	0.1519	0.9982	0.0028

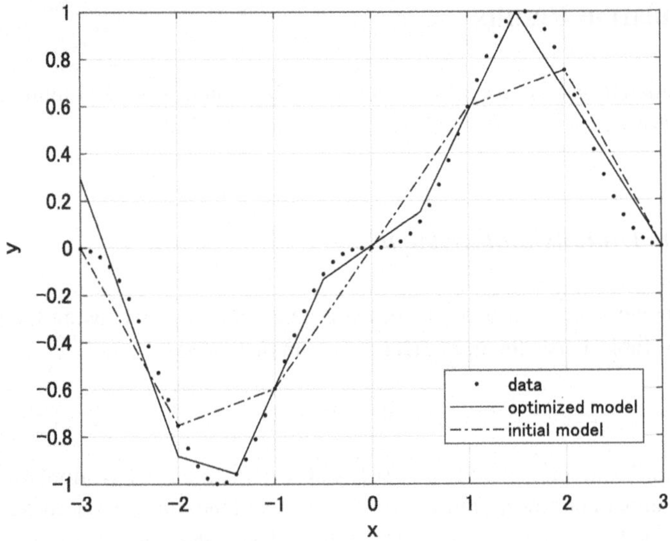

Fig. 1 Modeling result of example 1

4.2 Two-Dimensional System

Next, we consider numerical data ($N = 121 \times 121$) generated by the following equation:

$$\begin{pmatrix} y_1 \\ y_2 \end{pmatrix} = \begin{pmatrix} x_2 \\ -x_1 + (1 - x_1^2)x_2 \end{pmatrix} \tag{8}$$

Because $y_1 = x_2$ is a simple linear system, the PML model of y_2 is constructed from numerical data using the proposed algorithm. The learning parameters $\tau = 1$, $\alpha = 0.05$, and $\beta = 0.05$ are used as simulation parameters. Tables 3 and 4 list the initial and optimized vertex values. The initial vertex values, y_2, are generated by substituting initial values, (x_1, x_2), into (8). The computational time was 1.139 second. The modeling error of the PML model with the optimized vertex values is 0.0672. However, the error when using the initial vertex values is 0.2552.

x^+ in the PML model (2) is replaced with $\dot{x}$ to evaluate the modeling performance based on trajectories. In Fig. 2, the solid line indicates the original system (8). The dashed and dash-dotted lines are PML models using optimal and initial values of Table 2. Therefore, PML model using the proposed algorithm exhibits an excellent modeling performance for the training data.

Table 3 Initial vertex values of example 2

		x_2				
	y_2	–3	–1.5	0	1.5	3
x_1	–3	27	15	3	–9	–21
	–1.5	5.25	3.375	1.5	–0.375	–2.25
	0	–3	–1.5	0	1.5	3
	1.5	2.25	0.375	–1.5	–3.375	–5.25
	3	21	9	–3	–15	–27

Table 4 Optimal vertex values of example 2

		x_2				
	y_2	–3	–1.5	0	1.5	3
x_1	–3	25.8536	13.6964	3.2180	–8.1266	–19.4731
	–1.45	3.1118	3.2044	1.1958	0.2065	–0.6858
	0	–4.3844	–2.0840	0	2.2832	4.1775
	1.65	1.9419	–0.1220	–1.4748	–3.5351	–4.8852
	3	20.8179	9.0390	–3.0907	–14.8556	–27

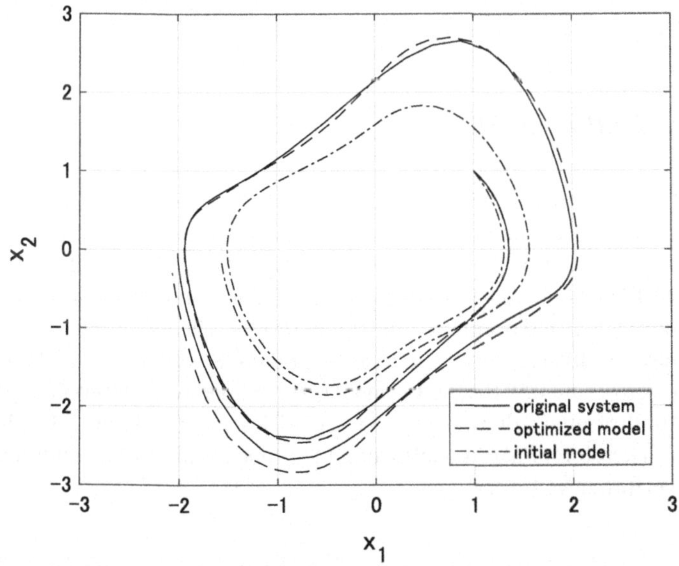

Fig. 2 Modeling results of example 2

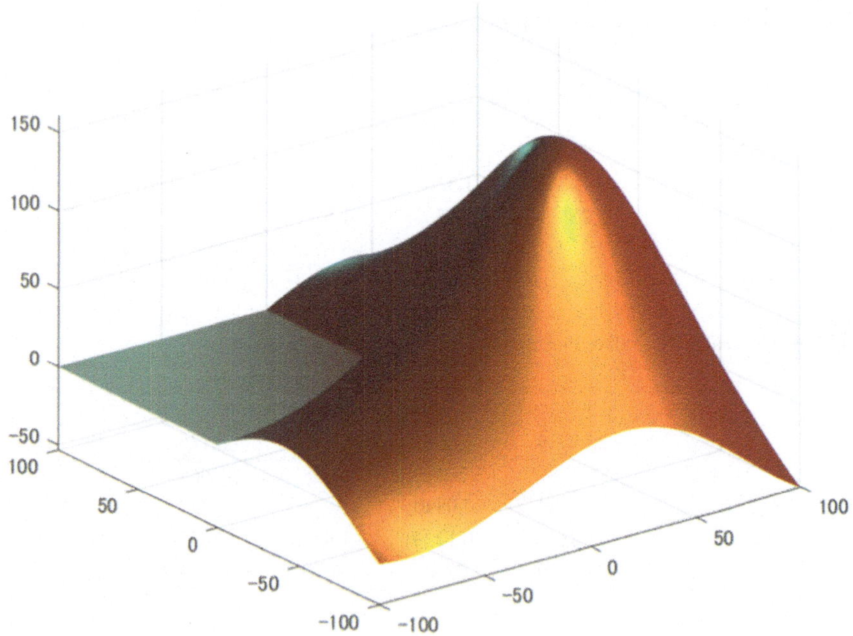

Fig. 3 MATLAB logo model

4.3 MATLAB Logo Model

We considered the numerical data of the MATLAB logo model. This model includes 201×201 training data and Fig. 3 reveals the three-dimensional graph using training data. Three PML models were constructed with 11×11, 21×21, and 41×41 vertices and the calculation times were 41.53, 124.9, and 268.4 seconds, respectively. Figures 3, 4, 5 and 6 show the modeling results. The modeling error decreases as the number of divided regions increases. In the cases of 11×11, 21×21, and 41×41, the errors are 0.9502, 0.056, and 0.0039, respectively. However, the modeling errors for the initial values of 11×11, 21×21, and 41×41 are 3.0711, 0.2045, and 0.0147. Therefore, PML models using the proposed algorithm exhibit excellent modeling performance for the data.

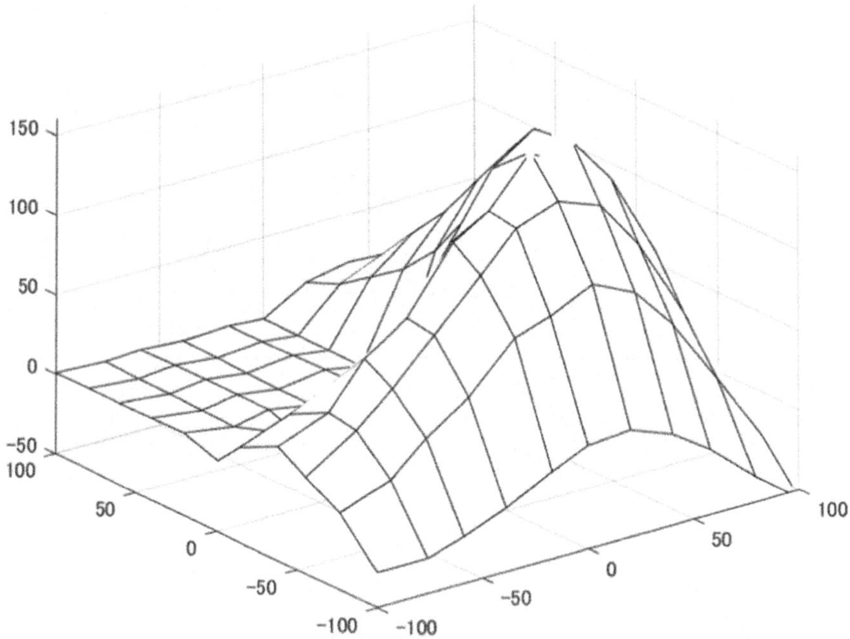

Fig. 4 PML model (11 × 11 vertices) using MATLAB logo data

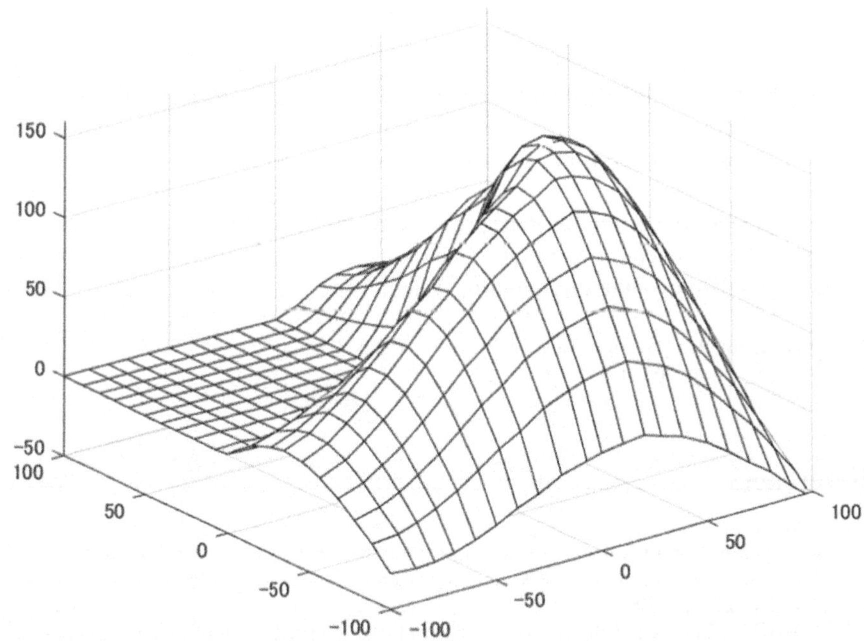

Fig. 5 PML model (21 × 21 vertices) using MATLAB logo data

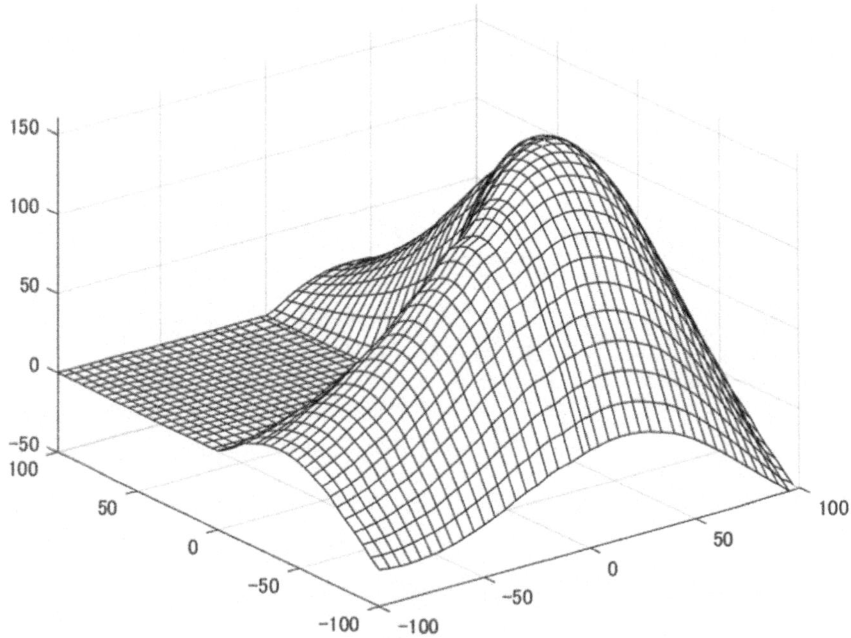

Fig. 6 PML model (41 × 41 vertices) using MATLAB logo data

5 Conclusion

This study entailed the development of a piecewise modeling method by numerical data. The piecewise model is represented as a rectangular region divided into state-space. The proposed algorithm finds simultaneously optimized vertex values and positions of segmented regions with minimum modeling errors. Three examples demonstrated the effectiveness of the proposed method using numerical simulations. In the future, we intend to consider noisy numerical data and apply the proposed modeling method to a nonlinear control system.

Acknowledgements This study was supported by KAKENHI Grant Number JP21K12051 of Japan Society for the Promotion of Science.

References

1. Bemporad A, Garulli A, Paoletti S, Vicino A (2005) A bounded-error approach to piecewise affine system identification. IEEE Trans Autom Control 50(10):1567–1580. https://doi.org/10.1109/TAC.2005.856667
2. Ho B, Kalman R (1996) Efficient construction of linear state variable models from input/output functions. At—Automatisierungstechnik 14:545–548 (1966). https://doi.org/10.1524/auto.1966.14.112.545

3. Ichida K, Yoshimoto F (1979) Spline functions and their applications. Kyoiku-shuppan. (in Japanese)
4. Ichihashi H, Watanabe T (1990) Learning control by fuzzy models using a simplified fuzzy reasoning. J Jpn Soc Fuzzy Theory Syst 2:429–437 (in Japanese)
5. Ljung L (1999) System identification: theory for the user. Prentice Hall information and system sciences series, Prentice Hall PTR
6. Nakada H, Takaba K, Katayama T (2005) Identification of piecewise affine systems based on statistical clustering technique. Automatica 41(5):905–913. https://doi.org/10.1016/j.automatica.2004.12.005
7. Åström KJ, Torsten B (1965) Numerical identification of linear dynamic systems from normal operating records. IFAC Proc Volum 2(2):96–111. https://doi.org/10.1016/S1474-6670(17)69024-4
8. Schoukens J, Ljung L (2019) Nonlinear system identification: a user-oriented road map. IEEE Control Syst Mag 39(6):28–99. https://doi.org/10.1109/MCS.2019.2938121
9. Sugeno M, Taniguchi T (2004) On improvement of stability conditions for continuous mamdani-like fuzzy systems. IEEE Trans Syst Man Cybern Part B (Cybernetics) 34(1):120–131. https://doi.org/10.1109/TSMCB.2003.809226
10. Taniguchi T, Sugeno M (2010) Stabilization of nonlinear systems with piecewise bilinear models derived from fuzzy if-then rules with singletons. In: International conference on fuzzy systems, pp 1–6. https://doi.org/10.1109/FUZZY.2010.5584807
11. Taniguchi T, Sugeno M (2019) Stabilization of discrete-time nonlinear systems using piecewise multi-linear control based on input-output feedback linearization. In: 2019 IEEE international conference on systems, man and cybernetics (SMC), pp 1582–1587. https://doi.org/10.1109/SMC.2019.8914016
12. Zadeh L (1956) On the identification problem. IRE Trans Circ Theory 3(4):277–281. https://doi.org/10.1109/TCT.1956.1086328
13. Zadeh LA (1962) From circuit theory to system theory. Proc IRE 50(5):856–865. https://doi.org/10.1109/JRPROC.1962.288302

Scaling of a Modular Production System (MPS) with Manufacturing Execution System (MES) and Multiple Agents

Paola Jéssica Llumiquinga, Manuel Alberto Armijos, William Paul Oñate, and Gustavo Javier Caiza

Abstract The digital transition currently taking place involves the existence of various studies to deliver solutions to digital industrial processes, and a systematic review of such studies has made evident a limitation with respect to those papers about implementation in the academy and confidential pilot tests developed by large industrial companies; all this situation decelerates the joint implementation of operation and information technologies (OT and IT) that enable scaling industrial automation and control systems (IACS). This paper implemented a manufacturing execution system (MES) to a Festo MPS-500 modular production system under the cores of the ISA-95 standard as a guide in data collection, production scheduling and product traceability; the stages of the production process are executed through multiple agents, in response to a mini batches order placed by a user through the internet, and the system also watches power consumption in case of no request, generating administrative and maintenance reports. The average latency metrics from the instant in which the user requests a mini-batch until the moment in which the plant starts the execution is 1.59% of the duration of the manufacturing process, indicating that the MES is not critical during the communication time with the server of the M2M (Machine to Machine) database.

Keywords MES · IT · OT · Cyber-physical systems · I4.0

P. J. Llumiquinga · M. A. Armijos · W. P. Oñate (✉) · G. J. Caiza
Universidad Politécnica Salesiana, Cuenca, Quito 170146, Ecuador
e-mail: wonate@ups.edu.ec

P. J. Llumiquinga
e-mail: jllumiquingac1@est.ups.edu.ec

M. A. Armijos
e-mail: marmijosv1@est.ups.edu.ec

G. J. Caiza
e-mail: gcaiza@ups.edu.ec

G. L. Conte and O. Sename (eds.), *Proceedings of the 11th International Conference on Mechatronics and Control Engineering*, Lecture Notes in Mechanical Engineering,
https://doi.org/10.1007/978-981-99-6523-6_6

1 Introduction

The manufacturing digital transformation requires that industrial automation and control systems (IACS) have available the technologies necessary for the adaptation of the manufacturing requirements [1] regarding customized products, smaller batches, unexpected events in the supply chain and interruptions; this situation gives rise to the optimization of manufacturing processes to achieve efficiency, flexibility, interoperability and savings [2].

To achieve competitiveness in the industrial manufacturing sector domestically and internationally it is necessary to implement smart manufacturing [3, 4], which requires cyber-physical systems (CPS) that enable information exchange between machines of industrial processes [5] and the company with the user; this makes possible to fulfill the request at any time and from any place, executing it in real-time from a manufacturing execution system (MES) [6]; however, this interaction requires resource management and planning through an enterprise resource planning (ERP) system [7].

As a consequence of the paradigm and the digital transition currently taking place, there are various studies about industrial IoT that have been implemented which, according to [8], corresponding to less than 10% of the total number of documents reviewed, indicating that most of the studies within this percentage are simulations and only 24% have been implemented. This indicates that there is classified information in the current literature [9–11], and the one found is limited; this is the case of [12], which designed an MES/MOM data model under the ISA-95 standard and its implementation in a smart production laboratory; however, the document does not show specific information about its development thus giving more emphasis to the design, whereby it intends to be a guide for future case studies. Similarly [13], shows how to develop a multiagent MES for a modular production system, with the use learning (ML) algorithm to analyze the trends in the behavior of the actuators and the number of parts produced; although this paper seems to include the implementation, from the literature it is seen a clear contribution in the design and not in the implementation, which represents a research niche for future works. On the other hand [14], proposes the use of the Industry 4.0 (I4.0) technology for a friction stir welding manufacturing process in the assembly of car parts at the level of a Computer Numerical Control (CNC) through a cloud-based MES, with real-time information exchange between the MES using a database and ERP, where the latter is the graphical interface for the user to place batch orders; as in the previous studies, it was observed that the document seems to contain coarse scientific dissemination; however, its information postulate regarding the practical execution of the system was not addressed in this study.

Based on the above, this paper intends to delve deeper into the case studies that have been implemented for MES with multiagent features, and develop an architecture for the communication between the manufacturing industry and the user, to place small batch orders in a modular production plant (MPS).

This document consists of Sect. 1 Introduction, Sect. 2 Methodology, Sect. 3 Analysis of Results and finally conclusions in Sect. 4.

2 Methodology

To scale an MPS to a system with I4.0 technology, it is necessary to construct a design of the communication architecture from the production plant to the user, for which it was decided to develop an MES based on the ISA 95 standard, which also consists of multiple agents, a database for processing production orders of small batches and a system that enables the user to place his/her order. In other words, this system is designed under a bidirectional communication architecture model in three levels [15]. The processing level is constituted by three stations of the MPS (conveyor, distribution and machining), that are responsible for producing two types of products; the first is a perforated red cylindrical part and the second is a silver cylindrical part without perforation. At this level, bilateral communication is carried out through the Message Queue Telemetry Transport (MQTT) protocol [16, 17]. Level two communicates with the lower level and with the upper level for controlling orders and information processing, respectively; this task is executed in a Raspberry (Node Red). Finally, the third level manages batch orders, corporate processes and manufacturing, through a platform constituted by a database (Firebase) and an interface for the final user, see Fig. 1.

2.1 *Architecture System*

As it has been specified in the knowledge of terms, a smart factory requires its functional archival operation to enable the interconnection, decentralization and interoperability of the systems, devices, online services and web platforms. For this purpose, it is appropriate to implement an MES that enables the execution of the production process in case of a request order; thus, it is required an architecture that enables the communication between the process, execution and product demand stages, for which it is necessary to have OT and IT.

Production Process. This stage has a modular production system (MPS) that for our case study was operated at the distribution, machining and conveyor stations, which are controlled by S7-300 S PLCs, as seen in Fig. 2.

Continuing with the manufacturing process in the MPS, the pallet moves forward through the conveyor to the machining station where, depending on the type of the product, one or two actions will be performed (verification and/or perforation); then, the stages that constitute this station are accomplished and thus the production process is finalized.

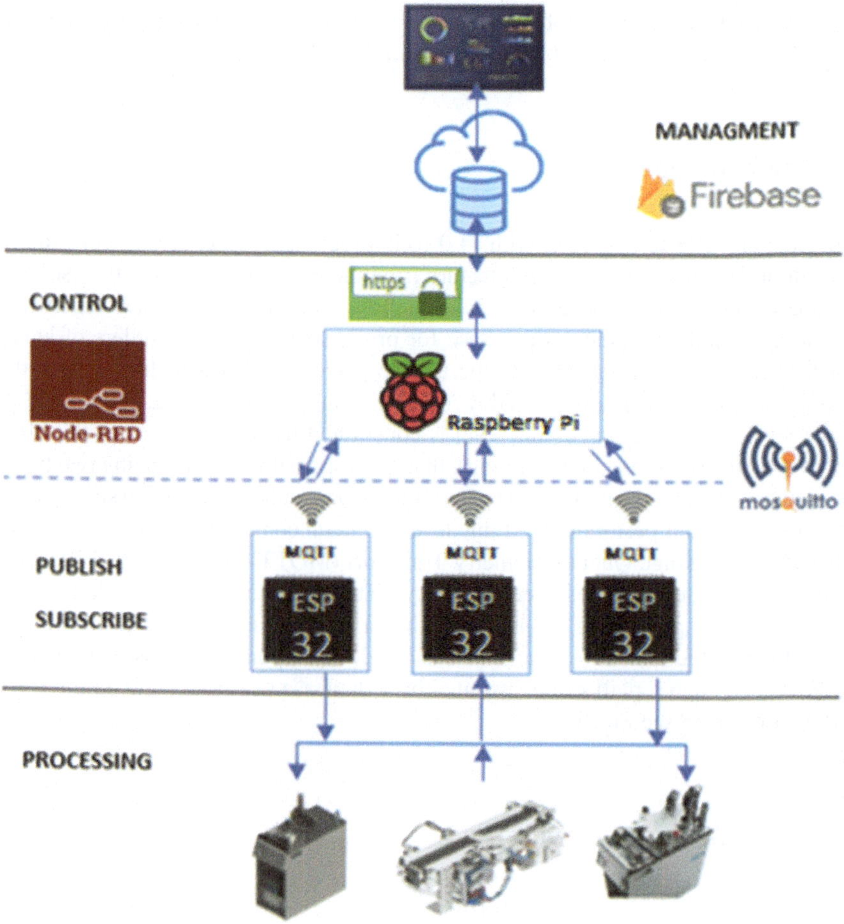

Fig. 1 MES architecture

As mentioned above, this station has a perforation module that consists of a permanent magnet DC motor, whose operating magnetic behavior is monitored in real-time. This information is sent to a database to generate an integrated report that facilitates maintenance access and planning by the operator.

On the other hand, the MPS stations described above are controlled using S7-300 PLCs, and they are in turn activated from the corresponding multiagent (ESP 32) through point-to-point communication.

Execution. A description of the ascending process within the communication architecture for a smart factory is now continued. This stage starts from the multiagent system where each of these devices (clients) executes the control command coming from the Raspberry Pi 3b (broker), with an MQTT light communication protocol [18] between these devices. For this case study and in this stage, the digital control

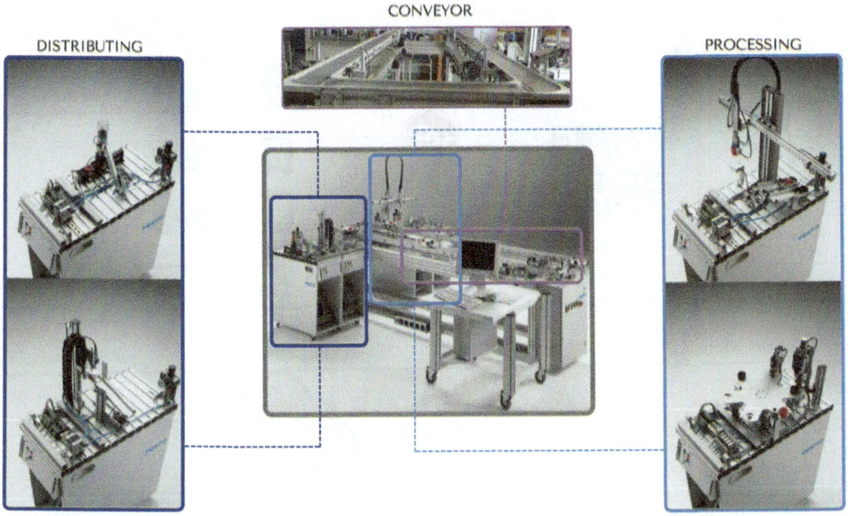

Fig. 2 Modular production system

by batches is in an administrative process that includes data collection, production scheduling and product traceability; however, there are other functionalities covered under the regulations and standards for a smart and flexible factory [2, 19]. To give way to what was mentioned above, a Node Red development tool was used for the connection between hardware devices and online services through application programming interfaces (API) with a bandwidth of 70 Mbps.

The block diagram of Fig. 3 shows that a user interacts in a digital environment (dashboard) to place a mini-batch order; this visual interface, which was developed in Node Red, also enables sending data in real time to a database (Google Firebase). Afterward, the MES schedules the production requesting the orders that were stored in the database, to execute them through the multiple agents, which are part of the production process. The 1880 port enables the Node-Red navigator for the communication between the MES and the database; likewise, the 1883 logical port is enabled from the raspberry for the communication with the multiple agents, with an established static IP address (172.17.140.9), as shown in Fig. 3.

From the process described above, it was developed programming using class diagrams, as shown in Fig. 4; starting from a logical sequence of specific actions by blocks, the flow initiates with the acquisition of information from the database in msg.payload type as simple JavaScript objects, considering attributes such as several parts ordered, their color and the date in which the order was placed, which will be distinguished using production rules configured from the Switch specialized node. In this manner, the numerical part of the msg.payload is extracted and the new message is resent through the corresponding outputs of this node to the inputs of other consecutive nodes, thus facilitating the recognition of the messages (A, B, D, and E) to group them in a vector of a Join node, which will be transmitted to a Q-gate

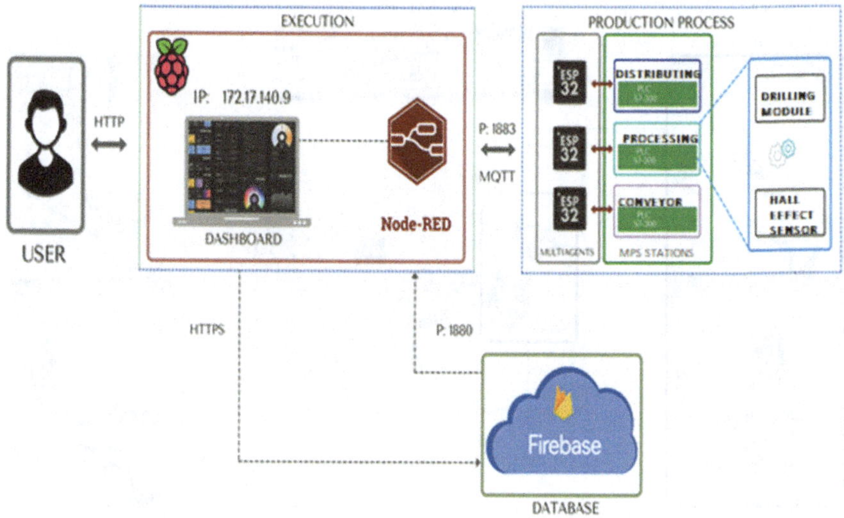

Fig. 3 Execution stage

specific node that adds the messages received to a queue, and places them in a list where they wait to be activated for the execution in a production chain.

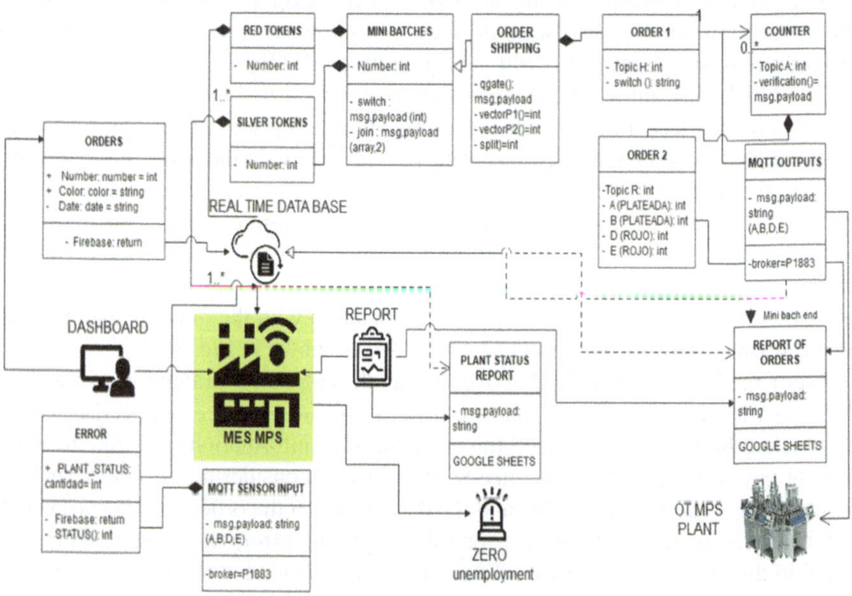

Fig. 4 Class architecture of the execution stage

Continuing to the shipping, an example is made of two orders in mini-batch placed from the dashboard, clicking on Start to activate the Q-gate node; in this manner, the message passes through two functions that determine the position of the orders (vector P1 and vector P2), which are worked individually and through a global function labeled by topics (Topic H and Topic R).

These features enable the multiple agents to execute the first order in an organized and instantaneous manner through the publication of an MQTT message from the broker. For the execution of the parts of the second order it is required the activation of a signal from the last production stage to the MES, indicates that the manufacturing of the first part has been completed; in this way, using a counter with Topic A and a comparator it is known the number of parts of the order that have been already shipped. Once the mini-batch has been shipped and finalized, the system is designed to zero nodes and execute the orders in the queue, and simultaneously it is placed in waiting time for a new batch; otherwise, if there is no order after this time has elapsed, the system stops feeding the Conveyor to save energy.

On the other hand, the MES also monitors the behavior of a traction machine, which through the subscription of a multiagent to the broker indicates the magnetic field status through MQTT. In addition, the system and using a status function it is evaluated the appropriate operating of the variable of interest, indicating the machining station status and also acting as a gateway for this information to be collected in the database.

The data generated and processed in the Execution stage Sect. 3.1.2 (mini-batches, orders, parts, monitoring of the machine status, date and time) are shown in a downloadable online report linked to the Google Sheets platform, thus enabling to visualize in real-time the number and color in the mini-batch orders, as well as the date and time of the status of an actuator of the machining station when an error has occurred.

The Demand for Products. While the Process Production and Execution stages are operational, it is executed the Demand of Products stage corresponds to the Firebase-Google database, where it is stored the data generated from the previous stages regarding the administrative process, which for this case study are data collection, production scheduling and plant status; it can be accessed from any place and at any instant [19]. The library Node-Red-Contrib-Firebase and the nodes corresponding to this action, firebase modify and firebase auth, respectively, were used to make it possible to send and request information, as observed in Fig. 5.

In the market there are available SQL databases (such as MySQL, Oracle and Access among the most popular ones) and no SQL databases (such as Redis, MongoDB, CouchDB, Cassandra, etc.). A real-time database provided by Firebase was used for this case study; due to its features, this platform offers a wide range of services for handling massive amounts of data, the synchronization of the information with different external applications, and simple design due to its JSON tree storage structure, and consequently its construction is not limited to a predetermined sheet but it enables that the developer customizes it according to the interests of the project.

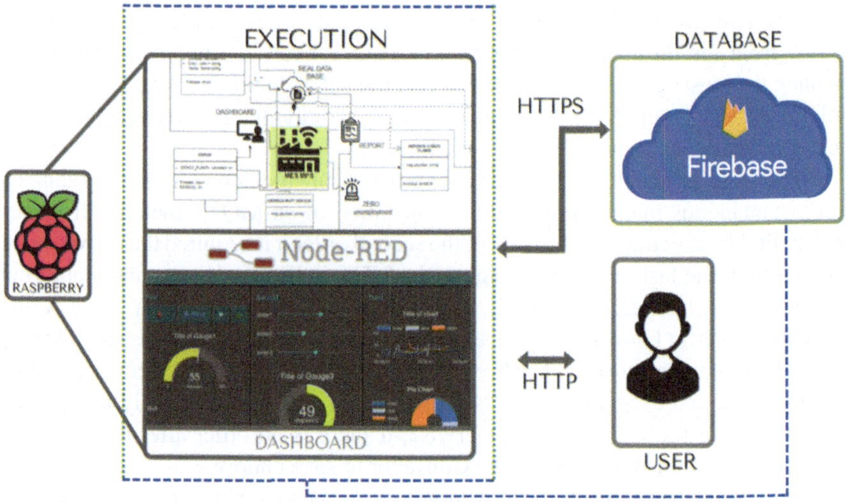

Fig. 5 Architecture of the communication between user—Database—MES

Taking into account what was previously mentioned, Fig. 6 presents a structural design of the database, which starts with the main collection (MAIN) that stores 3 documents (silver, red and sensor), the first two for production and the third for the plant status. In this manner, it is possible to define in each document attributes that would be desired to obtain in the future, such as number, date, status and value.

Taking into account the latency problems when using the cloud, Real Time Database was selected as the data storage alternative, because the MES is not regional with the location of the server nam-5 central (Iowa-USA).

Since the MES is bidirectional in its communication concerning the database, the tree storage structure was implemented at the first instance in Node Red, using the libraries and nodes for this case; in this manner it is reflected in a Firebase MAIN project. In this project, the rules for using a service limited in time and space, the authentication credentials for the account and the use of a URL were considered; the latter will be interpreted by the nodes that send and request information for the link between Firebase and Node Red. This process is repeated for the entire tree, i.e., collection, documents and attributes, as indicated in Fig. 6.

To generate the reports of production (administration) and plant status (maintenance), it is necessary to develop, in the MES visual development tool, a structural function for a data presentation sheet in Excel online, and also generate an access link from the Google Drive Google Forms service for the linkage, in which the attributes are modified at convenience for the automatic generation of a document that contains information of interest, visualized and downloadable from Google Sheets.

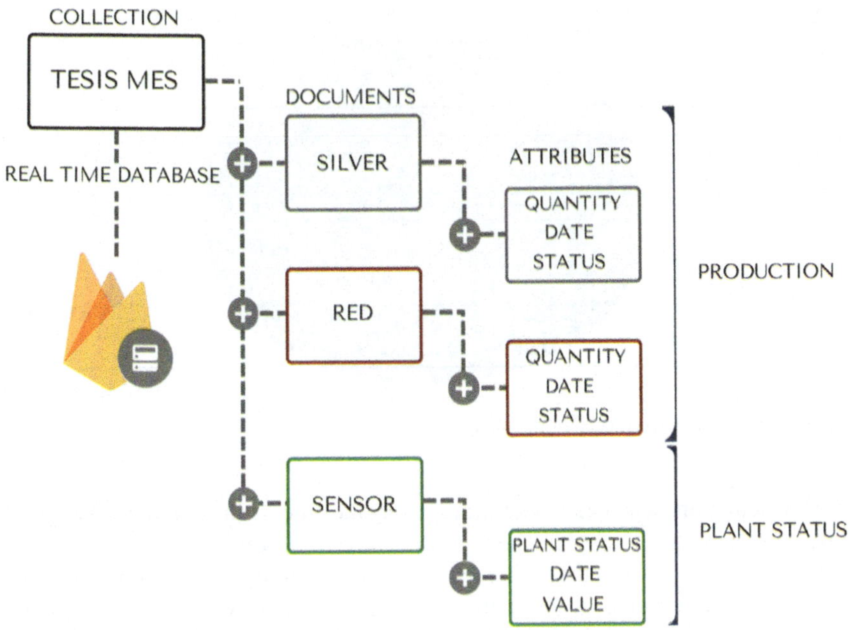

Fig. 6 Tree storage structure in firebase

3 Analysis and/or Interpretation of Results

Since the Festo MPS-500 has I3.0, it does not allow the user to interact with the production orders through the internet. Therefore, this study implemented OT and IT for scaling such plant to I4.0 improving production, energy consumption and competitiveness, through MES and multiple agents that execute the action of the S7-300 PLCs within the production line. Through a web interface, the user selects the product and executes the order, and in this manner the MES plans the production in conjunction with a database.

It should be known that the network infrastructure where the architecture shown in Fig. 1 was installed is not dedicated, and it is also mentioned that the production plant is located in the Universidad Politécnica Salesiana in Quito—Ecuador, and the database is in Iowa—USA; as a consequence, for analyzing the performance of the manufacturing plant scaled to I4.0, it is considered time metrics such as the plant response to an order by the user and the monitoring of the status of the motor magnetic field to the database, through Node Exporter, and therefore a test of mini-batch orders products and numbers was conducted, as it is observed in Fig. 7.

From a such figure it was evident that from a total of 20 orders, 17 orders were fulfilled successfully and without any problem. Nevertheless, the network was lost 3 times, causing a momentaneous production pause; despite this, the system resumed

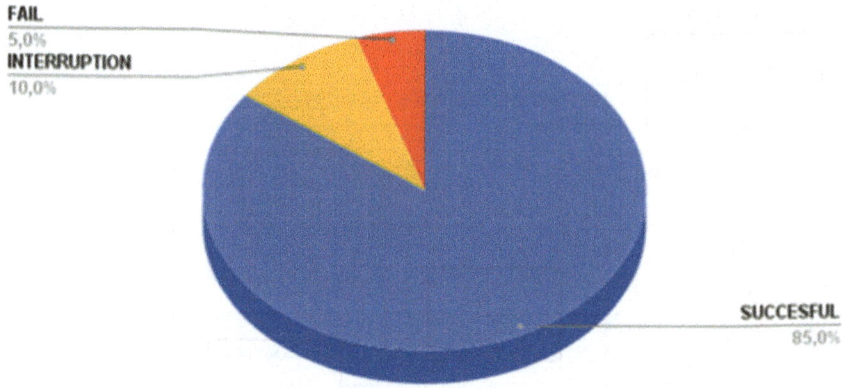

Fig. 7 Production performance of the plant

the production line where it was left, completing the process successfully; in contrast, it was not possible to complete the process on one occasion.

Figures 8 and 9 show the results of the plant response times to an order by the user and the monitoring of the status of the motor magnetic field in the database. To interpret the increase in Dashboard-Production time, it is important to remark that the system has a bidirectional MES-Database/Database-MES architecture for this process, as opposed to the Production-Dashboard where the information is sent unidirectionally from the MES to the database.

Regarding the plant operating status, the MES monitors every 500 [ms] the magnetic field of a permanent magnet motor that is in the machining station. To this magnetic field intentionally and randomly, a magnet was manually approached, and the values were scaled to 1 to identify certain anomalies; in other words, a comparison

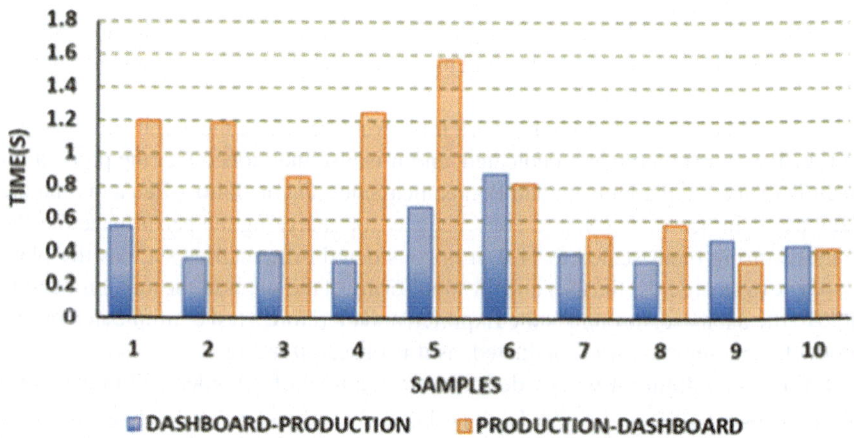

Fig. 8 Production response times

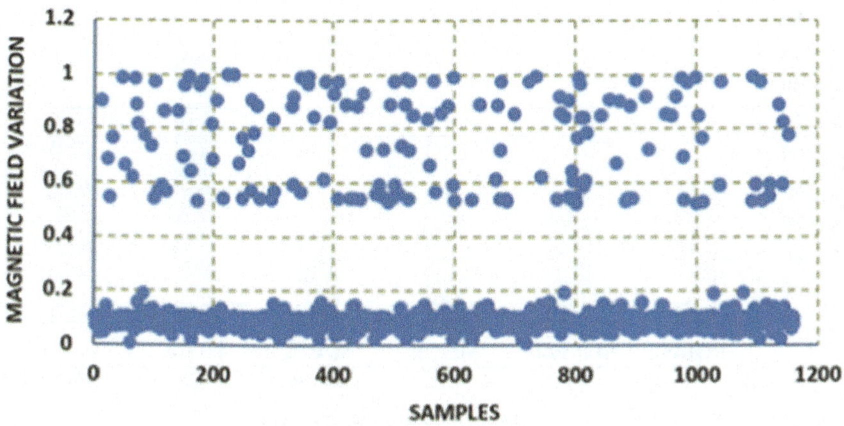

Fig. 9 Monitoring of the magnetic field of a DC motor (perforation of tokens)

was made concerning predetermined values when such field shows good behavior (values below 0.795) as indicated in Fig. 9.

Consequently, the data compared are stored in the database so that an operator analyzes the trends that involve the status of this actuator. Similarly, an administrative report was obtained which shows the product most demanded by the user, and the economic income and total sales per product; this test involved a total of 100 orders with values established: a cost of 0.90 USD for the red part and a cost of 0.75 USD for the silver part, as it is observed in Fig. 10. It is also mentioned that having all this information stored in a database, facilitates the analysis of the future behavior of the devices of interest through probabilistic and AI tools to prevent production stops due to maintenance; similarly, the production and sales information may be sent to the suppliers, so they have raw material in stock.

4 Conclusions

From the results obtained related to the scaling from I3.0 to I4.0, it was evident that the system operates acceptable manner even though there were limitations in the network infrastructure, which were specified in Sect. 4. This means that the MES successfully executed the manufacturing process 95% of the time, despite interruptions in the network infrastructure, thus demonstrating the need and importance of operating on a database typical of the characteristic cores of the ISA-95 standard, facilitating the interaction of the user with the production process.

The mean production time of a part (red or silver) is about 48 s, and the average latency metric from the instant when the user requests a mini-batch until the instant in which the plant starts the execution is 1.59% of the duration of the manufacturing process, indicating that this M2M (Machine to Machine) communication time is not

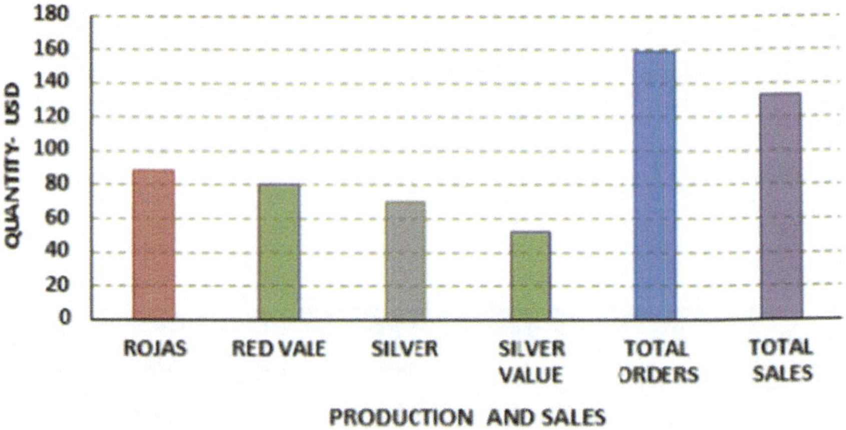

Fig. 10 Production and sales administrative report

critical. During the experimental tests to monitor the plant status, information relevant to the administration and operation activities was stored, and this data contributes to the manufacturing corporate decision making; however, the MES implemented in a Festo MPS-500 still poses already known but not yet implemented challenges that require the OT/IT synergy for managing stocks, computer systems, redundant systems such as databases in MES, virtual reality, digital twins, industrial communication networks that are time sensitive at the production level, etc.; under this situation, this study also intends to motivate with niches of systems implemented that have not been simulated, typical a manufacturing industry.

References

1. Berenice C, Manuel J, Guadalupe J, Pereyra FA, Osorio ML (2017) El Entorno de la Industria 4.0: Implicaciones y Perspectivas Futuras. Conciencia Tecnológica (54). https://www.redalyc.org/jatsRepo/944/94454631006/html/index.html
2. Chicaiza F, García CA, Castellanos EX, Sánchez C, Rosero C, Garcia Sanchez M (2018) Arquitectura flexible basada en ISA 88 para el diseño del diagrama de control de ejecución en aplicaciones distribuidas usando IEC 61499. Enfoque UTE 9(1):149–165. https://doi.org/10.29019/enfoqueute.v9n1.186
3. Luis J, Román D (n.d.) Industria 4.0: la transformación digital de la industria. http://coddii.org/wp-content/uploads/2016/10/Informe-CODDII-Industria-4.0.pdf
4. Banda H (2015) Impacto de la Manufactura Inteligente en la Industria y la Academia. https://www.researchgate.net/publication/283296307_Impacto_de_la_Manufactura_Inteligente_en_la_Industria_y_la_Academia
5. García CA, Castellanos EX, García MV (2019) Desarrollo de sistemas ciber-físicos de producción para Procesamiento por lotes usando normas IEC-61499 e ISA-88. Ingeniare Revista Chilena de Ingeniería 27(3):443–453. https://doi.org/10.4067/s0718-33052019000300443

6. Htun ZM, Borisovich FE (2019) Integrated production system using ERP and MES. In: 2019 2nd international conference of intelligent robotic and control engineering (IRCE). https://doi.org/10.1109/irce.2019.00014
7. Beric D, Havzi S, Lolic T, Simeunovic N, Stefanovic D (2020) Development of the MES software and integration with an existing ERP Software in industrial enterprise. In: 2020 19th international symposium INFOTEH-JAHORINA (INFOTEH). https://doi.org/10.1109/infoteh48170.2020.9066345
8. Asghari P, Rahmani AM, Javadi HHS (2018) Internet of things applications: a systematic review. Comput Netw. https://doi.org/10.1016/j.comnet.2018.12.008
9. Whitepaper (n.d.) Building a digitalization infrastructure for the process industry (2021). https://www.tttech-industrial.com/pdf/2021_12-TTTech-Industrial-Whitepaper-process-industry.pdf
10. Manager A, Gey G (n.d.) An Internet of intelligent things for more sustainable operations (2021). https://www.slb.com/-/media/files/software-integrated-solutions/industry-article/202104-worldoil-internet-of-intelligent-things-ia.ashx
11. Singh M, Makarychev G, Mustapha H, Voleti D, Akkurt R, Al Daghar K et al (2020) Machine learning assisted petrophysical logs quality control, editing and reconstruction. https://scholar.google.com/citations?view_op=view_citation&hl=en&user=TQIgQeIAAAAJ&citation_for_view=TQIgQeIAAAAJ:2osOgNQ5qMEC
12. Mantravadi S, Møller C, Chen LI, Schnyder R (2022) Design choices for next-generation IIoT-connected MES/MOM: an empirical study on smart factories. https://www.researchgate.net/publication/353850924_Design_choices_for_next-generation_IIoT-connected_MES MOM_An_empirical_study_on_smart_factories.
13. Mantravadi S, Li C, Møller C (2019) Multi-agent manufacturing execution system (MES): concept, architecture & ML algorithm for a smart factory case. In: Proceedings of the 21st international conference on enterprise information systems. https://doi.org/10.5220/0007768904770482
14. Adetunla A, Madonsela N (2021) Harnessing industry 4.0 technology to improve productivity in manufacturing industries: a case of CNC. https://www.researchgate.net/publication/353200869_Harnessing_Industry_40_Technology_to_Improve_Productivity_in_Manufacturing_Industries_a_Case_of_CNC_Based_Welding_Factory
15. Bogdanovs N, Krumins A, Belinskis R (2018) Design of wireless sensor networks for building management systems. In: 2018 22nd international conference electronics, pp 1–5. https://ortus.rtu.lv/science/en/publications/27957
16. Juwita PS, Fadhil R, Damayanti TN, Ramadan DN (2020) Smart parking management system using SSGA MQTT and real-time database. TELKOMNIKA (Telecommun Comput Electron Control) 18(3):1243. https://doi.org/10.12928/telkomnika.v18i3.14869
17. Porras Garzón N, Gómez Pérez HA (2020) Desarrollo de un sistema de comunicación basado en radio definido por software para un Robot Modular. Udistrital.edu.co. http://hdl.handle.net/11349/24317
18. Fernández Martínez EF, Ordóñez Narváez JA (2021) Desarrollo de una arquitectura Cloud Computing para comunicar dos estaciones del sistema de producción modular utilizando el protocolo MQTT. Ups.edu.ec. http://dspace.ups.edu.ec/handle/123456789/19958
19. León D-D, Junin G-T, Ludivina C-R, Raúl J (2017) Implementando objetos del estandar ANSI/ISA-95 con el patrón de diseño decorador "decorator." Articulo Revista de Prototipos Tecnológicos Marzo 3(7):20–26. https://www.ecorfan.org/spain/researchjournals/Prototipos_Tecnologicos/vol3num7/Revista_de_Prototipos_Tecnologicos_V3_N7_4.pdf; Estructura tu base de datos | Firebase Documentation (2021)

[illegible] Industry 4.0 [illegible]

[illegible] (2019) [illegible]

[illegible] (2019) [illegible]

[illegible] (2019) [illegible]

An Effective Methodology for Imbalanced Data Handling in Predictive Maintenance for Offset Printing

Alexandros S. Kalafatelis, Nikolaos Nomikos, Angelos Angelopoulos, Chris Trochoutsos, and Panagiotis Trakadas

Abstract The printing industry is one of the largest manufacturing industries in the world, being characterized by high production volumes, where continuous maintenance of machine performance is key. Predictive Maintenance (PdM) enables the use of a maintenance policy based on novel Machine Learning (ML) algorithms, in or-der to provide valuable insights for both diagnostics and prognostics. However, real-world data used for PdM model training are characterized by great class imbalances, as failure events have a significant lower rate of happening compared to the normal no failure operations. Furthermore, ML models that are subjected to imbalanced datasets, are prone to be highly biased while having misleading accuracy scores. This can prohibit systems to accurately predict machine failure, leading to excessive costs while affecting the safety of the workers. This work proposes a data sampling methodology for predictive maintenance algorithms used mainly in Offset Printing environments, aiming to improve model performance. Based on a historical dataset extracted by an Offset Printing manufacturer, a methodology consisting of multiple classification algorithms utilizing different sampling techniques (SMOTE,

A. S. Kalafatelis (✉) · N. Nomikos · A. Angelopoulos · P. Trakadas
Department of Port Management & Shipping, National and Kapodistrian University of Athens, 34400 Psachna, Evia, Greece
e-mail: akalafat@core.uoa.gr

N. Nomikos
e-mail: nomikosn@pms.uoa.gr

A. Angelopoulos
e-mail: a.angelopoulos@uoa.gr

P. Trakadas
e-mail: ptrakadas@pms.uoa.gr

C. Trochoutsos
Pressious Arvanitidis, 15232 Athens, Greece
e-mail: chtrox@pressious.com

G. L. Conte and O. Sename (eds.), *Proceedings of the 11th International Conference on Mechatronics and Control Engineering*, Lecture Notes in Mechanical Engineering,
https://doi.org/10.1007/978-981-99-6523-6_7

ADASYN, and RUS), was trained and evaluated using cross-validation. The evaluation outcomes demonstrated the ability of the proposed methodology to effectively handle data imbalances while significantly enhancing model performance, outperforming other state-of-the-art techniques.

Keywords Predictive maintenance · Machine learning · Industry 4.0 · Offset printing

1 Introduction

Predictive maintenance (PdM) has been gaining prominence recently in multidisciplinary sectors, enabling the use of a maintenance policy based on novel Machine Learning (ML) algorithms. In essence, PdM works by estimating and foreseeing failures in deteriorating systems around manufacturing environments, in order to optimize maintenance efforts [1].

The printing industry is one of the largest manufacturing industries in the world, having high production volumes, where continuous maintenance of machine performance is key. Possible breakdown events will automatically result in production stop, disturbing thus not only the production process, but also burdening financially the manufacturers. Offset Printing enables the production of large quantities, as the variable production costs are deemed small compared to the setup costs, thus having a greater risk in case of machine breakdown. Possible failures found in Offset Printing, include but are not limited to: (i) defective offset rubbers, (ii) wear of ink rollers, (iii) incorrect bending or damaged printing plates, (iv) insufficient pressures on the printing machines, (v) non-conformity issues in the sheet delivery unit, and (vi) random failures, which are found at every manufacturing environment [2].

According to Haarman et al. [3], maintenance procedures are shown to represent a total of 15–60% out of the total costs of operating of all manufacturing, thus showcasing the importance of a PdM solution. In detail, a PdM solution aims to not only prevent possible failures but to also optimize operations, affecting thus different aspects of manufacturing, including safety, product quality, reliability, and minimization of operational costs.

PdM data provide valuable insights for both diagnostics and prognostics information, enabling maintenance work to become proactive. ML assumes that data used for training and testing purposes are under the same feature space, having similar distribution and comparable proportion of training instances belonging to each class. However, this is not always the case in real world applications, where ML have to face complex challenges in which these assumptions are not always satisfied [4].

Furthermore, ML models that are subjected to imbalanced datasets, are prone to be highly biased while having misleading accuracy scores. This phenomenon can be attributed due to the lack of information coming from the minority class of a given dataset and to ML models in general, as they tend to classify every test sample into the majority class, in order to improve the accuracy metric [5, 6].

This phenomenon is predominated in cases where anomaly detection is of prime importance, such as in PdM, prohibiting the systems to accurately predict machine failure, leading not only to excessive costs for the manufacturers, but also possibly affecting the safety of the workers.

To mitigate this issue, sampling techniques such as under-sampling and oversampling are used either to create more instances of the minority class to increase its population or to minimize the data instances found on the majority class.

In this paper, the occurrence of machine failure is determined on a predictive maintenance dataset, implementing SMOTE, ADASYN and RUS methods to generate balanced datasets of machine failure instanced found in Offset Printing. The efficiency of the proposed oversampling and undersampling methodologies are analyzed with the help of various machine learning classifiers, with the aim to improve predictive maintenance accuracy scores.

This paper is structured as follows. In Sect. 2, we suggest the details of the utilized dataset and of the proposed methodology of handling imbalanced datasets, alongside with the classification algorithms. In Sect. 3, the experimental results used to assess the performance of the different sampling methods and of the classification models, are presented. Finally, in Sect. 4 the results are summarized and discussed.

2 Materials and Methods

2.1 Dataset Description

The original dataset consisted of features and labels based on historical measurements collected during a 4-month trial period (03/07/2022–31/10/2022) from Pressious Arvanitidis, an Offset Printing manufacturer based in Greece. Each of the collected parameters and features, follows the process of a particular printing order (i.e., from the sales department to the quality assessment department). The order and factory related characteristics used in this paper are presented in Table 1.

Table 2 summarized the descriptive statistics of the independent and dependent variables of the complete dataset.

2.2 Data Processing Methodology

Due to the high-class imbalance in the initial raw dataset regarding the failure events (containing only 145 events of some type of machine failure out of the 4205 total printing runs), data preprocessing was performed to facilitate the training and testing processes of the ML models with high-quality data.

Particularly, to avoid a scenario where a particular variance dominates the objective function of the learning algorithms (making it unable to learn from other features

Table 1 Parameters used for the training and testing procedures for the ML models

Parameter	Description
Unique order ID	Unique identifier varying from 1 to 10000
Quality	The requested end product paper type in a particular order. It is a categorical variable that takes values 'Velvet', 'Uncoated' and 'Illustration'
Quantity	The number of printing pieces requested in a particular order
Type	The outcome type of a particular order, taking values of 'Book', 'Poster' and 'Journal'
Color	The specific color requirements of an order. Categorical variable taking values between 'typical' 4-color printing, '4 + 1' color printing or 'grayscale' printing
Machine	The specific ID of the machine that a particular order was forwarded for printing, ranging from 1 to 5
Humidity	Water vapor relative to air temperature
Temperature	Air temperature at the factory ranging from 292 to 298 (K)
Tool Wear	The time required and used by a machine for each printing run
Failure	Indicates whether the machine has failed or not

Table 2 Parameters and attributes of the input and target variables

Parameter	Mean	Standard deviation	Minimum	Maximum
Unique Order ID	2969	1214.02327	867	5071
Quality	1.600238	0.761683	1	3
Quantity	2331.809750	1319.670624	206	9956
Color	3.680856	0.947442	1	5
Machine	2.676100	1.337525	1	5
Humidity	55.007498	3.391050	45.070	69.940000
Temperature	294.327795	1.041168	292	300.010
Tool wear	7.772699	4.398902	0.686667	33.186667
Failure	0.034483	0.182487	0	1

correctly as expected), data scaling was performed initially, using the Log Transformation methodology. The method was used as it enables data measurements to become more symmetric to a normal distribution. After the scaling the dataset was divided into a training set (80%) and test set (20%) (Step 1).

Furthermore, high data dimensionality has shown to have a direct effect on classification accuracy, increasing the rate of misclassification and thus reducing the overall accuracy of a classification algorithm. Therefore, dimensionality reduction was also performed using Principal Component Analysis (PCA). Specifically, the dataset PCA enables the conversion of correlated features found in the high dimensional space into a series of uncorrelated features in the low dimensional space, that depict the linear combination of existing variables, and for that reason it has become a necessity before applying any data sampling approach [7] (Step 2).

To effectively deal with the class imbalance, three different sampling techniques were employed, namely, Random UnderSampling (RUS) [8], Synthetic Minority Oversampling Technique (SMOTE) [9] and the Adaptive Synthetic sampling approach (ADASYN) [10] (Step 3).

These techniques operate in a feature space aiming either to under-sample the majority class data or oversample the minority one. On the one hand, under-sampling techniques such as RUS, are used to improve imbalance levels of the classes to the desired target, by reducing the number of majority instances. However, the removal of instances from the majority class is performed without replacement, meaning that useful information might be permanently lost. In addition, due to the randomized nature of RUS, an unclear decision boundary may be resulted, affecting classifiers performance [11].

On the other hand, over-sampling approaches intent to improve imbalance levels of the classes to the desired target, by generating synthetic instances and adding them to the minority class. Unlike approaches such as random oversampling, SMOTE generates artificial instances in the minority class, based on the feature space, rather than the data space, considering linear combinations between existing minority samples. Moreover, derived from SMOTE, the ADASYN approach gives different weights to different minority samples of a given dataset, while it automatically determines the number of samples required to produce in order to achieve data balance.

The aforementioned methodology is depicted in Fig. 1.

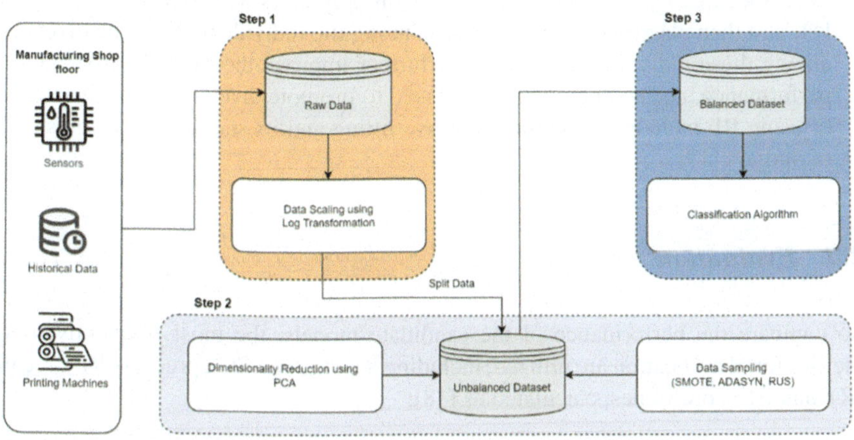

Fig. 1 Proposed methodology for imbalanced data handling in predictive maintenance

2.3 Machine Learning Models

To create the proposed framework, stratified 5-fold cross validation was used for all the experiments in this study. The base ML models were trained, using the scikit learn package [12], including:

- Logistic regression (LR) is a standard probabilistic statistical classification model that has been extensively used for classification problems across disciplines. Different from linear regression, logistic regression analyzes the relationship between multiple independent features and estimates the probability of occurrence of an event, by fitting the data onto a logistic curve. LR is affected by outliers, which greatly skews parameter estimation, reducing classification performance [13].
- k-Nearest Neighbors (kNN) [14] enables a low-power computational classification through the identification of the nearest neighbors given by a query example and using those neighbors to determine the class of the query [15].
- Decision Tree (DT) is a learner which repeatedly splits the dataset according to a cost criterion that maximizes the separation of the data, resulting in tree-like branches. In detail, the algorithm attempts to select the most important features to split branches and iterate through a given feature space. Compared with the other machine learning methods, decision trees have the key advantage, that are not characterized as black-box models and can be easily expressed as rules [16].
- Random Forest (RF) algorithms fall under the broad umbrella of ensemble learning methods. The key principle underlying the algorithm is the decision tree. Specifically, every data instance is initially classified by every individual DT, and then classified by a consensus among the individual DTs. The diversity among these individual DTs can thus further improve the overall classification performance, and so bagging is introduced to promote diversity. The advantages of using RF include its robustness to overfitting and its stability in the presence of outliers [17].

2.4 Evaluation

To compare the performance of the candidate models, the most frequently used metrics for classification are utilized, including accuracy (ACC), precision (P), recall (R), and F1-score values, calculated as [18]:

$$\mathrm{ACC} = (\mathrm{TP} + \mathrm{TN})/(\mathrm{TP} + \mathrm{FN} + \mathrm{TN} + \mathrm{FP}) \tag{1}$$

$$P = \mathrm{TP}/(\mathrm{TP} + \mathrm{FP}) \tag{2}$$

$$R = \mathrm{TP}/(\mathrm{TP} + \mathrm{FN}) \tag{3}$$

$$F1 = (2 \cdot \text{Precision} \cdot \text{Recall})/(\text{Precision} + \text{Recall}) \quad (4)$$

3 Results

Table 3 showcases the overall performance comparison between the different classification algorithms, each utilizing different sampling methodologies with the aim of achieving higher accuracy and f1-scores, for more accurate classification of machine failures in the field of Offset Printing.

In detail, the experiment results demonstrate that both Random Forest and Decision Trees algorithms performed significantly better than the rest of the base models, while Logistic Regression performed the least accurate scores. Moreover, both SMOTE and ADASYN sampling methods, showed to improve classification accuracy throughout the models, while Under-sampling had the least effect on improving classification accuracy.

Furthermore, as showcased in Fig. 2, the implementation of SMOTE and ADASYN indicated similar results, with models under SMOTE slightly outperforming the rest of the methods using ADASYN.

Table 3 Overall performance evaluation of classification algorithms under different sampling methodologies

Model	Data sampling method	Accuracy	Precision	Recall	F1-score
Logistic regression	Under sampling	0.468304	0.924963	0.468304	0.609441
	SMOTE	0.563391	0.929364	0.563391	0.692212
	ADASYN	0.543582	0.928011	0.543582	0.675927
k-nearest neighbors	Under sampling	0.541204	0.933273	0.541204	0.673145
	SMOTE	0.751189	0.925741	0.751189	0.827805
	ADASYN	0.755151	0.924929	0.755151	0.830226
Decision trees	Under sampling	0.496830	0.939275	0.496830	0.633158
	SMOTE	0.841521	0.930260	0.841521	0.882692
	ADASYN	0.828051	0.929606	0.828051	0.874811
Random forest	Under sampling	0.516640	0.931664	0.516640	0.652170
	SMOTE	**0.882726**	**0.930196**	**0.882726**	**0.905511**
	ADASYN	0.874802	0.928736	0.874802	0.900716

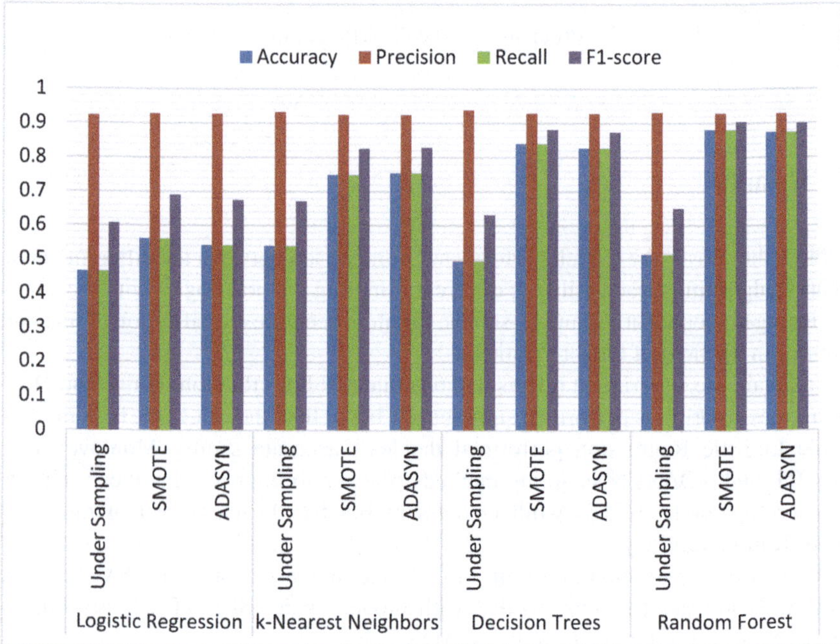

Fig. 2 Performance comparison of the proposed algorithms

4 Conclusions

Predictive Maintenance systems are utilized to predict trends, behavior patterns, and correlations by ML models in order to anticipate pending machine failures in a proactive manner, thus avoiding downtime and production stop. Machine maintenance has therefore attained critical importance for manufacturing industries such as the ones found in Offset Printing, due to the current growth in complexity of the manufacturing ecosystems.

In this study, we proposed a data sampling methodology for predictive maintenance algorithms for Offset Printing environments, which aims to effectively balance data classes and improve the performance of PdM models to accurately identify the minority classes using binary classification. The methodology consisting by the SMOTE, ADASYN, and RUS techniques, and the classification algorithms (DT, LR, KNN, RF), was generated based on a dataset from an Offset Printing company.

Overall, the results of this study indicate that the proposed methodology effectively handles data imbalances while enhancing model performance in classification accuracy, by outperforming other state-of-the-art techniques. Moreover, to the best of our knowledge, ours is the first study to explore PdM systems and data handling approaches for the Offset Printing domain.

Finally, in future work, the proposed methodology can be further extended for multi-class classifications, as well as the evaluation of further ML and DL techniques.

Acknowledgements This work has been partially supported by the PDS project, under the open call of the AI REGIO (Regions and Digital Innovation Hubs alliance for AI-driven digital transformation of European Manufacturing SMEs) project, funded by the European Commission under Grant Agreement number 952003 through the Horizon 2020 program (https://www.airegio-project.eu/) and by the ICOS (Towards a functional continuum operating system) project, funded by the European Commission under Grant Agreement number 101070177 through the Horizon 2020 program (https://www.icos-project.eu/).

References

1. Selcuk S (2016) Predictive maintenance, its implementation and latest trends. Proc Inst Mech Eng Part B: J Eng Manuf 231(9):1670–1679
2. Bălan E, Berculescu L, Răcheru RG, Piţigoi DV, Adăscăliţa L (2021) Preventive maintenance features specific to offset printing machines. In: MATEC web of conferences, vol 343
3. Haarman M, Mulders M, Vassiliadis C (2021) Predictive maintenance 4.0: predict the unpredictable. PwC Mainnov 4(30)
4. Gazzah S, Hechkel A, Amara NEB (2015) A hybrid sampling method for imbalanced data. In: 2015 IEEE 12th international multi-conference on systems, signals & devices (SSD15), pp 1–6
5. Kalafatelis A, Panagos K, Giannopoulos AE, Spantideas ST, Kapsalis NC, Touloupou M, Kapassa E, Katelaris L, Christodoulou P, Christodoulou, Trakadas P (2021) ISLAND: an interlinked semantically-enriched blockchain data framework. In: International conference on the economics of grids, clouds, systems, and services, pp 207–214
6. Spantideas ST, Giannopoulos AE, Kapsalis NC, Angelopoulos A, Voliotis S, Trakadas P (2022) Towards zero-defect manufacturing: machine selection through unsupervised learning in the printing industry. In: Proceedings of the workshop of I-ESA, Valencia, SP
7. Kotekani SS, Velchamy I (2020) An effective data sampling procedure for imbalanced data learning on health insurance fraud detection. J Comput Inf Technol 28(4):269–285
8. Saripuddin M, Suliman A, Syarmila Sameon S, Jorgensen BN (2021) Random undersampling on imbalance time series data for anomaly detection. In: 2021 the 4th international conference on machine learning and machine intelligence, pp 151–156
9. Chawla NV, Bowyer KW, Hall LO, Kegelmeyer WP (2002) SMOTE: synthetic minority oversampling technique. J Artif Intell Res 16:321–357
10. He H, Bai Y, Garcia EA, Li S (2008) ADASYN: adaptive synthetic sampling approach for imbalanced learning. In: 2008 IEEE international joint conference on neural networks (IEEE world congress on computational intelligence), pp 1322–1328
11. Zuech R, Hancock J, Khoshgoftaar TM (2021) Detecting web attacks using random undersampling and ensemble learners. J Big Data 8(1):1–20
12. scikit-learn. https://scikit-learn.org/stable/. Accessed 17 Nov 2022
13. Feng J, Xu H, Mannor S, Yan S (2014) Robust logistic regression and classification. Adv Neural Inf Process Syst 27
14. Zhang Z (2016) Introduction to machine learning: k-nearest neighbors. Ann Transl Med 4(11):218
15. Angelopoulos A, Giannopoulos AE, Kapsalis NC, Spantideas ST, Sarakis L, Voliotis S, Trakadas P (2021) Impact of classifiers to drift detection method: a comparison. In: International conference on engineering applications of neural networks, pp 399–410

16. Myles AJ, Feudale RN, Liu Y, Woody NA, Brown SD (2004) An introduction to decision tree modeling. J Chemom Soc 18(6):275–285
17. Angelopoulos A, Giannopoulos A, Spantideas S, Kapsalis N, Trochoutsos C, Voliotis S, Trakadas P (2022) Allocating orders to printing machines for defect minimization: a comparative machine learning approach. In: IFIP international conference on artificial intelligence applications and innovations, pp 79–88
18. Flach P, Kull M (2015) Precision-recall-gain curves: PR analysis done right. Adv Neural Inf Process Syst 28

Design and Experimental Validation of RL-Based Decision-Making System for Autonomous Vehicles

Ana María Gómez Ruiz, Hussam Atoui, and Olivier Sename

Abstract In autonomous driving, different Reinforcement Learning (RL) methods have been implemented to deal with different challenges. One of its advantages is the capability to deal with unexpected situations after an adequate trained environment. The inclusion of RL algorithms is considered as a solution for autonomous driving called "agent" that gathers the environmental information and acts according to this from one state to the next one. This paper proposes a solution for a specific environment that is trained with Deep RL and then is tested in simulation and in on experimental platform.

Keywords Deep reinforcement learning · Decision making · Autonomous vehicles

1 Introduction

Autonomous driving systems have raised a considerable interest in the last decades for several reasons. Initially, it can decrease the majority of lethal accidents that are caused by distracted drivers which will create safer roads. More than 90% of reported traffic accidents are the outcome of human error and caused by issues related to the acquisition of visual information as debated in [10]. Nevertheless, sophisticated

A. M. Gómez Ruiz (✉) · O. Sename
University Grenoble Alpes, CNRS, Grenoble INP, GIPSA-Lab, 38000 Grenoble, France
e-mail: ana.gomez@univ-grenoble-alpes.fr

O. Sename
e-mail: olivier.sename@gipsa-lab.grenoble-inp.fr

H. Atoui
Valeo, Driving Assistance Research (DAR) Team, 94000 Créteil, France
e-mail: hussam.atoui@valeo.com

G. L. Conte and O. Sename (eds.), *Proceedings of the 11th International Conference on Mechatronics and Control Engineering*, Lecture Notes in Mechanical Engineering,
https://doi.org/10.1007/978-981-99-6523-6_8

autonomous driving can decrease accidents caused by human errors, can redirect driving time into more productive ends and it can lower operating costs per mile finding optimal paths to destination.

The autonomous driving system have been under fast development in the recent years and different approaches have been implemented. Common modules to design autonomous systems include localization, perception, decision making (path planning) and dynamics control [13]. The main task of the environment localization and perception module is to extract useful features from the surroundings and locate the vehicle in the track to establish spatial and temporal relationships among the vehicle [3]. Identifying objects in the road, pedestrians, bicycles among others is classification ability that has raised a great interest with Machine Learning algorithms specially with supervised learning. To get this information from the vehicle environment, the module relies on different kinds of perception sensors such as cameras, radar and lasers [14].

The trajectory planning module aims to plan different longitudinal and lateral vehicle maneuvers which might include lane changing, braking, lane following and obstacle avoidance. There are existing methods that rely on traditional classical planners or machine learning methods. An alternative approach to the classical planners and supervised learning methods is Reinforcement Learning. This framework works on the principle of maximizing reward for a particular action at a given state [5]. RL is the theory of an agent that learns optimal behavior through interaction with its environment. With the aid of Deep RL techniques it is possible to use the benefits of deep learning in conjunction with RL to learn optimal behavior from high dimensional inputs to action outputs as discussed in [11]. In this paper, Actor-Critic methods are used to combine value-based and policy-based algorithms to sample efficiency and stability being effective in high dimensional and stochastic actions.

The general objective of the project is to build, integrate and test different modules of perception and control for a scaled autonomous vehicle in the Robot Operating System (ROS) framework. The car is grouped by Engineers at Gipsa-Lab. Previous work has been made in the vehicle such as identification of the model's vehicle, and its actuators along with the main connections on ROS2. In addition, the design and implementation of robust controllers for the vehicle lateral dynamics using different approaches has been made.

The following work aims to design the decision making module based on deep RL approach. The vehicle must avoid collisions, achieve high driving efficiency by taking an optimal path, and execute smooth maneuvers without veering off the track while maintaining the center-line of a two-lane race track. The RL model is trained in simulation with a Deep Q-Network, and is then validated and tested in an experimental scenario with the scaled RC car.

This paper is organized in five sections. The Sect. 2 aims to explain the theory and main components of the RL Algorithms that are used and the Actor Critic Approach. The Sect. 3 explains the implementation and training of the RL model with some simulation results. The Sect. 2 shows the validation and experimental results, and the Sect. 5 are the conclusions and final remarks of the work made.

2 Reinforcement Learning for Autonomous Driving

For autonomous driving, different Machine Learning (ML) methods have been implemented to deal with different challenges. Some of these algorithms have raised great interest because of the capability to deal with unexpected situations after an adequate trained on a large set of sample data. One of the biggest challenges with ML algorithms for autonomous driving is when considering the vehicle in an open context environment to train the model with all possible scenarios in the real world. The variety of context that could happen are infinite and the companies leading this field must solve it by collecting a big amount of data and validating system operation based on the collected data to ensure that a self-driving car has already learned all possible scenarios and with safety scenarios for each case [1]. The inclusion of RL algorithms is being considered as a solution for the car called agent that gathers the environmental information and acts according to this from one state to the next one.

The general idea for implementing RL algorithms is to take the most important aspects of a learning agent that is interacting with its environment to reach a goal. The agent must be capable to perceive the state of the environment described as observation and it must be able to take actions that affect its state; refer to [17]. This agent also has a reward according to the state of the environment and the objective is to obtain the highest value for the sum of rewards over the long run.

The RL algorithms are considered closed-loop because the actions taken by the agent influence its later inputs. As a difference with ML algorithms, the agent is not guided to which action to take but instead to discover which actions will yield to the most reward by exploring them out. In the most complicated cases, actions may affect not only the immediate reward, but also the next situations and all the subsequent rewards. Such characteristic of not having a direct instruction on what action to take, and the consequences of actions are the most important features of the reinforcement learning problems [12]. The goal is to find a sequence of inputs that drive a dynamical system to maximize some objective, beginning with minimal knowledge of how the system responds to inputs.

2.1 *Elements of Reinforcement Learning*

In order to explain the elements of the RL algorithm some definitions for the interaction to achieve a goal will be explained. The learner and decision-maker is called the *agent*, the ego vehicle. The agent interacts with what is called the *environment* which includes everything outside the agent, i.e., the racetrack, the obstacles and the surrounding vehicles. The agent takes an action which results in a change in the environment. This interaction is received by the agent as a *state* which includes information about coordinates and/or speed of other vehicles, features of the road, among others. Refer to Fig. 1 to visualize the connection between these components.

The main subelements of RL algorithms are:

- Policy: Is a mapping from perceived states of the environment to actions to be taken in those actions. It is sufficient to determine the behavior, policies may be stochastic.
- Reward: The objective of the agent is to maximize the cumulative reward received over the long run. This value depends on the agents current action and the current state of the agent's environment at any time. The only way the agent can influence the reward signal is through its actions, which can have a direct effect on the total reward, or an indirect effect through changing the environment's state. The policy may be changed to select the action that will be followed by a higher reward on that situation in the future.
- Value function: Specifies what is good in the long run defined as episode. The value of a state can be described as the total amount of reward an agent can expect to accumulate over the future, starting from that state. Whereas rewards determine the immediate, intrinsic desirability of environmental states, values indicate the long-term desirability of states after taking into account the states that are likely to follow, and the rewards available in those states.
- Model: Is a representation of the behavior of the environment. When an action is made given a state the model might predict the resultant next state and reward due to this action. The model is used for planning and to consider possible future situations before they actually happen.

2.2 *Reinforcement Learning theory*

The interaction between the agent and the environment occurs at a sequence of discrete time steps t in which it receives some representation of the environment's state $S_t \in \mathcal{S}$ in the $\mathcal{S}$ set of possible states, and it selects an action $A_t \in \mathcal{A}(S_t)$ where $\mathcal{A}(S_t)$ is the set of actions available in state S_t. One time step later, in part as consequence of its action, the agent receives a numerical reward $R_{t+1} \in \mathcal{R} \subset \mathbb{R}$ and finds itself in a new state S_{t+1}. The Fig. 1 represents the agent-environment interaction. At each time step, the agent implements a mapping from states to probabilities of selecting each possible action. This mapping is called the agent's policy and is denoted π_t, where $\pi_t(a \mid s)$ is the probability that $A_t = a$ if $S_t = s$. Reinforcement learning methods specify how the agent changes its policy as a result of its experience. The agent's goal, roughly speaking, is to maximize the total amount of reward it receives over the long run [12].

An agent can increase the long-term reward by exploiting knowledge learned about the discounted sum of expected future rewards of different state-action pairs. The learning agent has to exploit what it already knows in order to obtain rewards, but it also has to explore the unknown in order to make better action selections in the future [2].

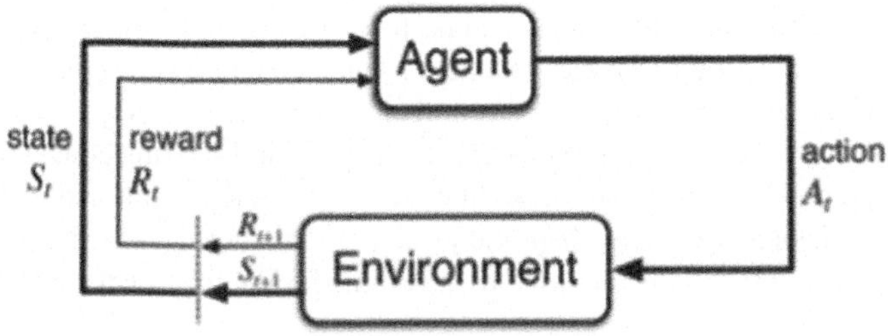

Fig. 1 The agent-environment interaction in RL. Image taken from [12]

For some stochastic control problems when the models for sequential decision making outcomes are uncertain, Markov Decision Processes (MDP) are used. The MDP model consists of decision epochs, states $\boldsymbol{S}$, actions $\boldsymbol{A}$, rewards $\boldsymbol{R}$, and transition probabilities $\boldsymbol{T}$; a tuple $< S, A, T, R >$. Choosing an action a in a state s generates a reward $R(s,a)$ and determines the state at the next decision epoch s' through a transition probability function $T(s, a, s')$. Policies are instructions of which action to choose under any occurrence at every future decision. The agent look for policies which are optimal [7]. The mathematical representation of the policy which is a mapping from the state space to a probability over the set of actions, and $\pi_t(a \mid s)$ represents the probability of choosing action a at state s. The goal is to find the optimal policy π^* at time k, defined as:

$$\pi^* = \underset{\pi}{\text{argmax}}\, \mathbb{E}_\pi\{\sum_{k=0}^{H-1} \gamma^k R(s_k, a_k) \mid s_0 = s\} := \underset{\pi}{\text{argmax}}\, V_\pi(s) \tag{1}$$

where γ is the discount factor that controls how an agent consider future rewards. When γ is low the agent will maximize short term rewards, on the contrary with high values of γ the agent will try to maximize rewards over a longer time frame. The Eq. (1) represents the highest expected sum of discounted rewards ([16]) in a time horizon H in the MDP. From the models directly, RL agents may learn value function estimates, policies and/or environment. Finding a policy π that maximizes the expected discounted sum of rewards over trajectories in the state space is what solving a RL task means.

2.3 *Reinforcement Learning Components for Autonomous Driving*

Some of the most important elements of the RL model are the actions, the state, the observations and rewards.

Actions The actions that the vehicle can perform are driven by the acceleration and the steering control of the vehicle. The actions are considered discrete for the agent to decide which distinct action to perform from a finite action set.

The *DiscreteMetaAction* type adds a layer of speed and steering controllers on top of the continuous low-level control, so that the ego-vehicle can automatically follow the road at a desired velocity. Then, the available meta-actions consist in changing the target lane and speed that are used as set points for the low-level controllers. The actions are listed as:

- 0: Lane left
- 1: IDLE
- 2: Lane right
- 3: Faster
- 4: Slower.

State The state of the vehicle, also named as observations, contains information of the agent and the vehicles around it. The *KinematicObservation* is the default of the library, this is an array of size $nObs\ x\ nF$ where n is the number of nearby vehicles and F is a set of features such as curvature, x, y, v_x, v_y. The number of vehicles n is constant and configured initially by the environment, so that the observation has a fixed size. The curvature of the track has been included as a the inverse of the lookahead radius ($\frac{1}{r}$) after several attempts of training the model. Its inclusion is an improvement to consider the approaching curve so that the agent can decrease the speed when getting into a pronounced curve that is 3m in front so it can keep the lane center trajectory.

$$\left[\tfrac{1}{r}\ x\ y\ v_x\ v_y\right]^{\top} \tag{2}$$

Rewards The final element to be defined are the rewards, the choice of an appropriate reward function yields realistic optimal driving behavior. A reward for collision, zero speed, lane centering and high speed has been defined. The total reward in every step will be determined by the sum of each condition. R_{coll} is the reward if it collides being -10 if it does and 0 if it does not. R_{stop} is the reward given if the vehicle stops, is 0 if the vehicle has some speed and -10 if it stops. R_{lc} is the reward given for lane centering, is maximum when the vehicle is in the center of the lane and it decreases proportionally when it moves away from the center lane as in Eq. (4). Finally, R_{hs} is the high speed reward and its value is a function of the speed of the vehicle as in 4. The total reward R_{total} is given by Eq. (3) and the final tuning of the rewards which resulted on the best simulation results is given in Table 1

$$R_{total} = R_{coll} + R_{stop} + R_{lc} + R_{hs} \tag{3}$$

$$R_{lc} = \frac{1}{1 + r_{lc} * lat\ error^2} - 0.5 \tag{4}$$

Table 1 Values of rewards of the RL model

Reward	Value
Collision r_c	–10
Zero speed r_{v0}	–10
Lane centering r_{lc}	1
High speed r_{hs}	0.7

r_{lc} is the weight of the lane centering, and $lat\ error$ is the difference between the reference trajectory and the position of the vehicle, the second term of Eq. (4) is also tuned. The values of speed are also discrete and could take 6 different values between 0 and $1.3m/s$ as shown in Eq. (5), when the speed is at the maximum, then the reward will be r_{hs}, if it decreases then the reward will decrease proportionally as the range of speed of the vehicle.

$$R_{hs} = r_{hs} \frac{index_v}{index_{v\max}} \tag{5}$$

After the R_{total} is obtained it is normalized between 0 and 1, and it becomes an input of the RL model.

2.4 *Actor Critic Approach*

Actor-critic methods are hybrid methods that combine value-based and policy-based algorithms. One actor is the one that selects the actions and this is the policy-structure. After an action is made by the 'actor', the estimated value function evaluates the action and this is known as the 'critic'. The value-based methods are model-free Temporal Difference (TD), are methods that can learn directly from raw experience without a model of the environment's dynamics and learn estimates of the utility of individual state-action pairs represented in Eq. (6) [15]. This scalar signal is the sole output of the critic and drives all learning in both actor and critic, as shown in Fig. 2.

$$Q_\pi(s, a) = \mathbb{E}_\pi \left\{ \sum_{k=0}^{H-1} \gamma^k R(s_k, a_k) \mid s_0 = s, a_0 = a \right\} \tag{6}$$

Q-learning will learn (near) optimal state-action values provided a big number of samples are obtained for each pair. Agents implementing Q-learning update their Q values according to the update rule of Eq. (7):

$$Q(s, a) \leftarrow Q(s, a) + \alpha \left[r + \gamma \max_{\alpha' \in A} Q(s', a') - Q(s, a) \right] \tag{7}$$

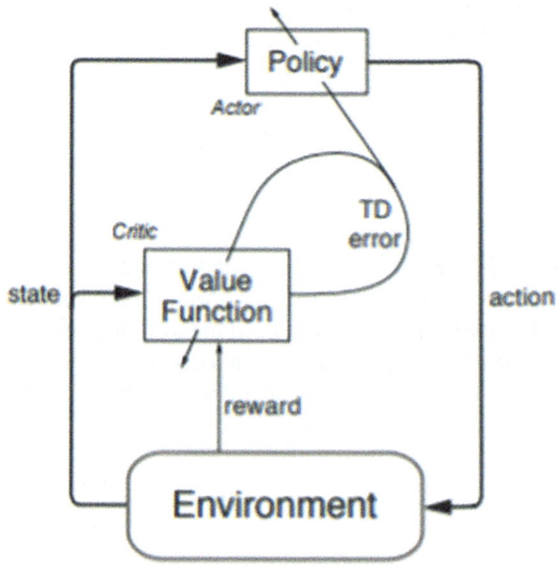

Fig. 2 The actor-critic architecture. Image taken from [12]

where $Q(s, a)$ is an estimate of the utility of selecting action a in state s; α is the learning rate which controls the degree to which Q values are updated at each time step [15].

The policy-based methods aim to estimate the optimal policy directly, and the value is a secondary. Typically, a policy π_θ is parameterized as a neural network. Policy gradient methods use gradient descent to estimate the parameters of the policy that maximize the expected reward. The result can be a stochastic policy where actions are selected by sampling, or a deterministic policy. When selecting actions, exploration is performed by adding noise to the actor policy. To stabilize learning a replay buffer is used to minimize data correlation. A separate actor-critic specific target network is also used. Normal Q-learning is adapted with a restricted number of discrete actions the optimal Q-value and optimal action as Q^* and a^*.

$$Q^*(s, a) = \max_{\pi} Q_\pi(s, a) \ , \ a^* = \text{argmax}\, a Q^*(s, a) \tag{8}$$

By correcting the Q-values towards the optimal values using the chosen action, the policy is updated towards the optimal action proposition. Thus two separate networks work at estimating Q^* and π^*.

3 Training and Testing of the Model

The aim of the project is to drive a scaled vehicle on a racetrack without veering off track or crashing, and reaching an optimal speed to finish a loop as fast as possible. Initially, the implemented solution has been developed in an existing framework named *highway-env* (github library), which is an open source Python library with a collection of different environments for autonomous driving and tactical decision-making tasks. This tool has been modified to create a new environment with specific dimensions of the track and the car for the particular environment of the vehicle and available space of the experimental room at GIPSA-Lab, in the next chapter the details of the scaled vehicle will be discussed. After setting the vehicle behavior and the environment, a RL model is used to estimate the action in every step of the trajectory given to the agent.

3.1 Vehicle Behavior

Some of the vehicle parameters of the vehicle are presented in Table 2 with the dimensions of the scaled vehicle. The motion of the vehicle is represented by the modified bicycle model shown in Fig. 3. The vehicle kinematics are presented by the following Eq. [6]:

$$\dot{x} = v\cos(\psi + \beta) \tag{9}$$

$$\dot{y} = v\sin(\psi + \beta) \tag{10}$$

$$\dot{v} = a \tag{11}$$

$$\dot{\psi} = \frac{v}{l}\sin(\beta) \tag{12}$$

$$\beta = \tan^{-1}\left(\frac{1}{2}\tan\delta\right) \tag{13}$$

where (x, y) is the vehicle position; v is the forward speed; ψ is heading angle; a is the acceleration command; β is the slip angle at the center of gravity; and δ is the front wheel angle used as a steering command. Its state is propagated depending on

Table 2 Car parameters of the scaled vehicle

Parameter	Unit	Value
Mass	kg	1.34
Length	m	0.174
Width	m	0.0870
Wheel radius	m	0.0650

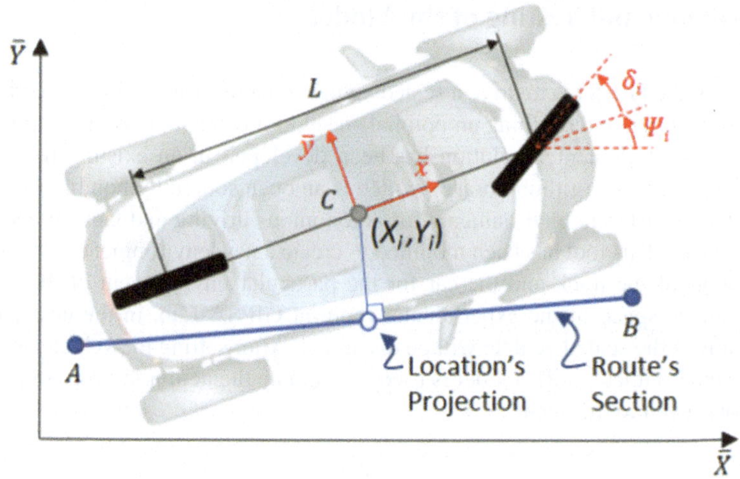

Fig. 3 Lateral vehicle dynamics. Image taken from [4]

the steering and acceleration actions. For the vehicle dynamics, the two degrees of freedom are represented by the vehicle lateral position y and the vehicle yaw angle Ψ. The vehicle lateral position is measured along the lateral axis of the vehicle to the point C which is the center of rotation of the vehicle. The vehicle yaw angle Ψ is measured with respect to the global X axis. The longitudinal velocity of the vehicle at the center of gravity is denoted by V_x. The Eq. (14) represents the lateral translational motion of the vehicle and Eq. (15) represents the moment balance about the z axis.

$$m(\ddot{y} + \dot{\Psi} V_x) = F_{yf} + F_{yr} \tag{14}$$

where F_{yf} and F_{yr} are lateral tire force of the front and rear wheels, respectively.

$$I_z \ddot{\Psi} = l_f F_{yf} - l_r F_{yr} \tag{15}$$

where l_f and l_r are the distances of the front tire and the rear tire respectively from the center of gravity of the vehicle [8].

The controlled vehicle is a low-level controller, allowing to track a given target speed and follow a target lane. The longitudinal controller is a simple proportional controller as shown in the Eq. (16).

$$a = K_p\,(v_r - v) \tag{16}$$

The lateral controller is a simple proportional-derivative controller, combined with some non-linearities that invert those of the kinematics model. The position and heading control are shown in Eqs. (18)–(21), respectively.

$$v_{lat,r} = -K_{p,lat}\Delta_{lat} \tag{17}$$

$$\Delta\psi_r = \arcsin\left(\frac{v_{lat,r}}{v}\right) \tag{18}$$

where Δ_{lat} is the lateral position of the vehicle with respect to the lane center-line; $v_{lat,r}$ is the lateral velocity command and $\Delta\psi_r$ is a heading variation to apply the lateral velocity command.

$$\psi_r = \psi_L + \Delta\psi_r \tag{19}$$

$$\dot{\psi}_r = K_{p,\psi}\left(\psi_r - \psi\right) \tag{20}$$

$$\delta = \arcsin\left(\frac{1}{2}\frac{l}{v}\dot{\psi}_r\right) \tag{21}$$

where ψ_L is the lane heading (at some lookahead position to anticipate turns); ψ_r is the target heading to follow the lane heading and position; $\dot{\psi}_r$ is the yaw rate command; δ is the front wheels angle control; and $K_{p,lat}$ and $K_{p,\psi}$ are the position and heading control gains.

3.2 *Environment Setup*

The Fig. 4 shows the racetrack that has been used to train the model. Several factors are considered to build the environment, such as two lanes, straight segments, pronounced curves and obstacles in the road, which increase the complexity of the vehicle performance.

Obstacles The ego-vehicle, in green, is surrounded by other vehicles that have speed zero, which can be considered as objects or obstacles for this scenario to decrease the implementation complexity but for future works it is expected to threat as vehicles with different speeds.

Road The dimensions of the track are constrained by the experimental room in GIPSA-Lab with an available space of 4 m × 4 m. The maximum distance in the horizontal axis is 3.6 and 2.8 m in the vertical one. Finally, the track has been built with the union of 11 segments as union of straight lines and segments of different radius circles, the radius of each segment is an important characteristic that is included in the RL model and will be explained in the next section.

Training Procedure The components mentioned in the previous section enter a deep neural network which will estimate the action of the car in every step of the trajectory and can be trained from a Stable Baselines3 (SB3) github library that is a set of reliable implementations of reinforcement learning algorithms in PyTorch.

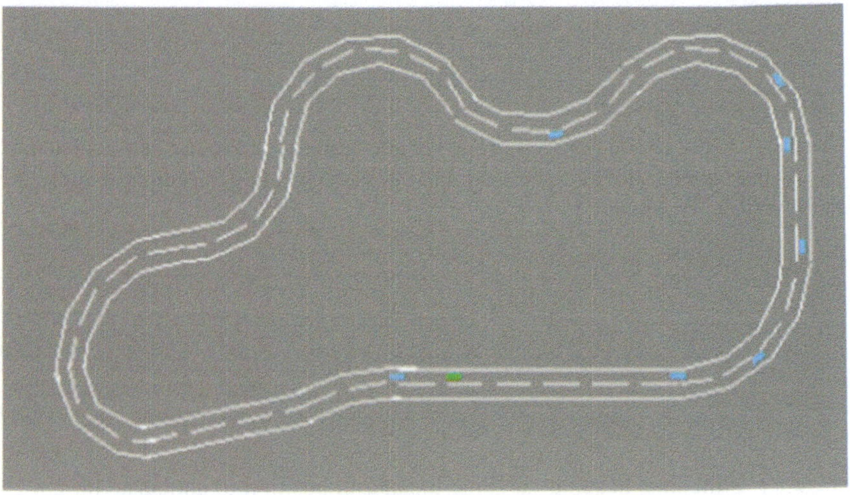

Fig. 4 Racetrack based on the highway-env library

Table 3 Parameters for training

Parameter	Value
Learning rate	0.0005
Discount factor	0.8
Exploration fraction	0.8
Total timesteps	60.000

This RL model is considered as an episodic domain that may terminate after a fixed number of time steps, or when an agent reaches a specified goal state. Also, the implemented policy is a Multilayer Perceptron (MLP) that consist of biased neurons arranged in layers, connected by weighted connections. 2 layers of 64 nodes have been used. Their effectiveness depends on finding the optimal weights and biases that reduce the classification error [9]. Some parameters of the training are displayed in the Table 3.

3.3 Training Results

After the combination of the vehicle behavior, the environment and the RL components, the training model has been performed on a Google Colab service that requires no setup to use, while providing access free of charge to computing resources including GPUs. The total training time is 5.5 h and the model output is obtained as a .zip extension for later use in the experimental results.

After training, the results can be reproduced for one episode by the states, observations, actions and rewards at every step. The duration of one episode is 140 s and

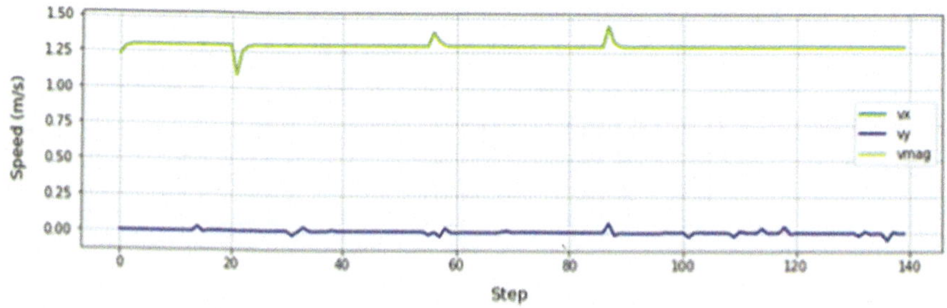

Fig. 5 Speed behavior of the vehicle in one episode

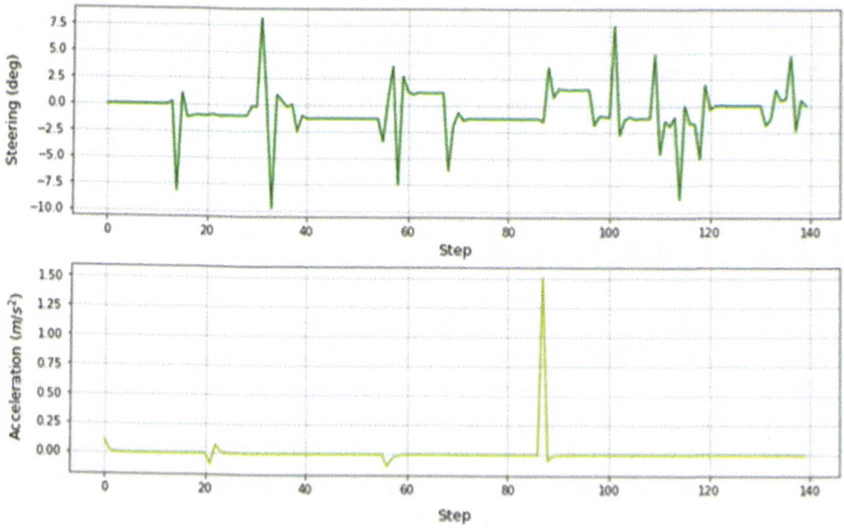

Fig. 6 Control variables of the vehicle in one episode

the frequency is of 2 actions per second. On the other hand, about the behavior of the vehicle, the available speed values are [0 , 0.26, 0.52, 0.78, 1.04, 1.3] m/s, and the speed profile is presented in Fig. 5for the magnitude, lateral and longitudinal values. Here the lateral speed is lower than 0.1 m/s and the magnitude is very similar to the longitudinal speed. The vehicle tried to complete a loop closer to the maximum available speed, here the functionality of the high speed reward is shown.

Additionally, the control variables are the steering angle and the acceleration. The steering angle that is limited between [–15 , 15]°. This limit affected the model and restricted the vehicle to behave more conservatively. The longitudinal acceleration is also restricted between [–2, 2] m/s^2 to avoid aggressive speed changes. The results are shown in Fig. 6.

Lastly, the actions and the output of the RL model is shown in Fig. 7. The predominant action is to accelerate but it also changes lane when facing obstacles and

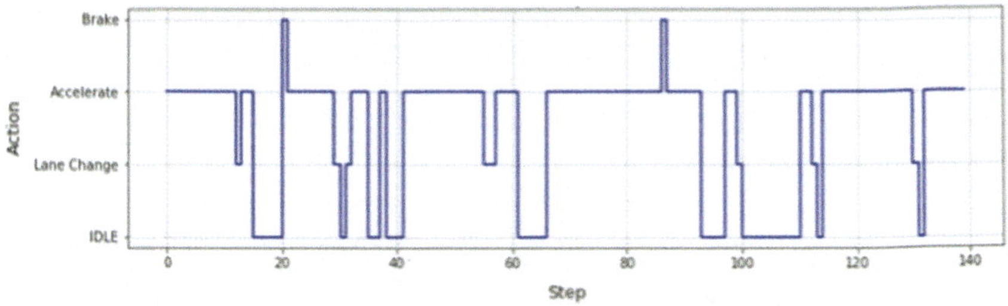

Fig. 7 Actions taken by the agent in one episode

in some pronounced curves. Because the behavior is predominant by the controllers, if the action would be to accelerate and if the speed limit is reached, the action can be 'ignored' and the vehicle kept an IDLE action, which for future works it would be preferable for the vehicle to take the available actions and not all of them.

4 Experimental Validation

A new environment is created with specific dimensions within the space of the experimental room, likewise, applied on the scaled vehicle in GIPSA-Lab. Figure 8 shows the experimental scenario where the validation of the results have been carried out, a two-lanes track is displayed as a reference. Table 4 shows the sensors and actuators of the vehicle enumerated in Fig. 9. In addition to the car components, there are high resolution cameras from which the position of the vehicle is measured with high accuracy. This sensor information of the road and the car is required to build the ROS2 architecture, that includes the perception, planning, decision making, and control nodes.

In the ROS2 decision making node, the observations have been programmed with the WiFi communication between the sensors and the computer. The RL model is trained in a track with a higher complexity than the one tested in this experiment. The decision making node includes the file model.zip containing the model, in which

Fig. 8 Experimental scenario setup at GIPSA-Lab

Table 4 RC car components

N	Type	Functionality
1	Switch	Switching car on-off
2	8 mm qualisys super-spherical	Captured by vicon tracker
3	Arduino RP 2040	Micro-controller of the vehicle
4	Spur gears	Increase torque given by BLDC
5	Elastic wheel	2 Rear wheel of the vehicle
6	ACCU NI-MH 3000	Supply power battery
7	MG996R servo motor	Steering actuator
8	Elastic wheel	2 Front wheels of the vehicle
9	BLDC-A2212/13T	Throttle actuator

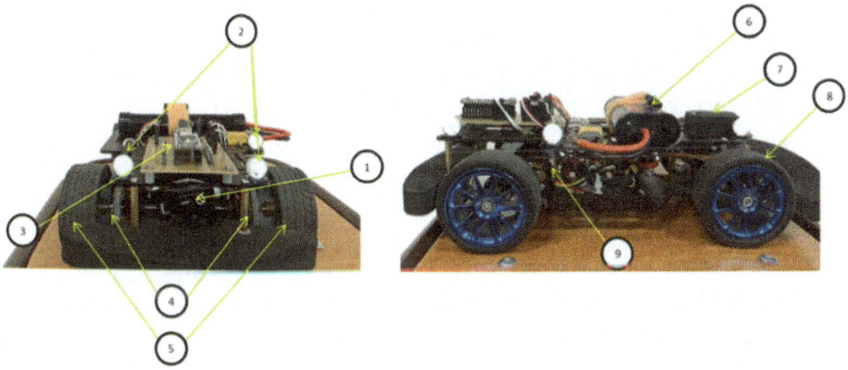

Fig. 9 Front and side pictures of the car

given the observation of the environment to predict the next discrete action to perform. Several scenarios have been tested with different obstacle positions. The first scenario with no obstacles is shown in Fig. 10, the vehicle can maintain the lane where it started. In this figure two loops are displayed with a different starting point of the vehicle in a direction counter clockwise.

The second scenario with one obstacle, the vehicle changes the lane before the curvature of the obstacle to avoid the crash, as shown in Fig. 11. The obstacle is displayed as the black square. The lane change is occurred between 4000 and 4200 ms, shown in orange in the figure.

The third scenario includes two obstacles, see Fig. 12, in which the vehicle changes the lane in advance to avoid collision. The times where the change lane action is performed are between 2000–2200 and 3200–3500 ms. One interesting observation is when the obstacle is located at the end of the curvature, the vehicle has the tendency

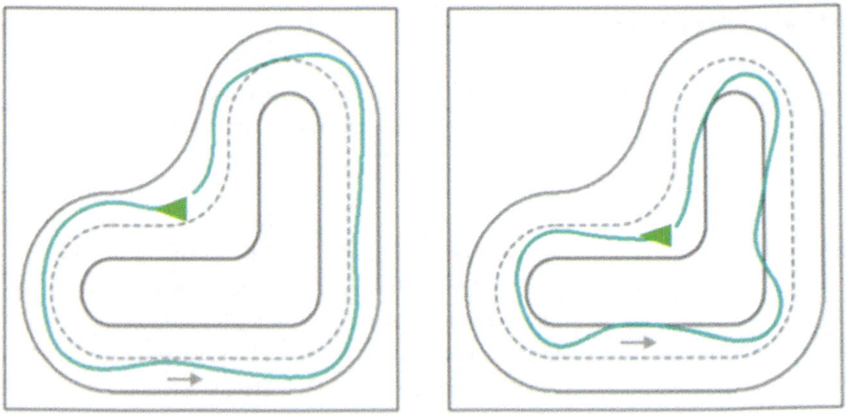

Fig. 10 Experimental results with no obstacles

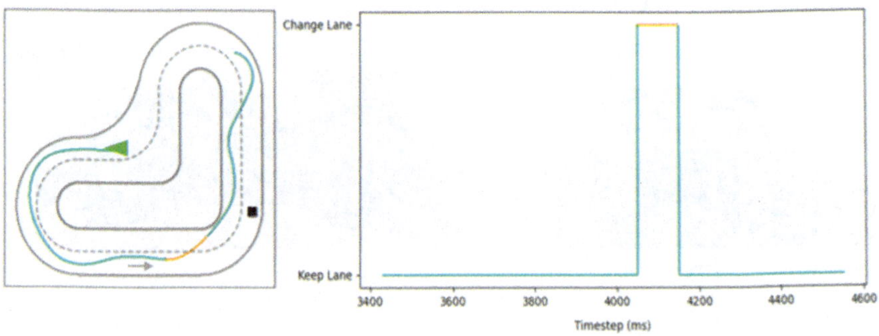

Fig. 11 Experimental results with one obstacle

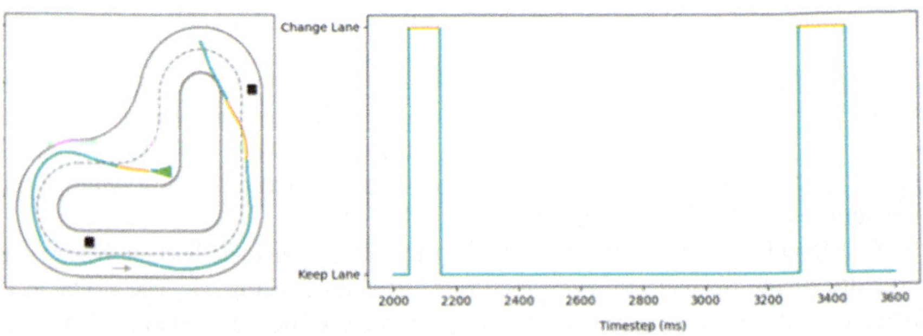

Fig. 12 Experimental results with two obstacles

to change lane in advance earlier than when the obstacle is on the straight path. This behavior has been observed in more scenarios that might be explained after the inclusion of the curvature as one attribute of the state in the training.

5 Conclusion

The performance validation of the RL model has been presented in simulation and in experimental tests. The results showed that scaled vehicle avoided obstacles, achieved high driving efficiency by taking an optimal path and executing maneuvers without veering off the track while maintaining the center-line of the two lane racetrack. Some remarks about the training model are improvement of the results obtained by including the curvature of the next segment in the track and also the influence of the rewards affect drastically the results, here one solution has been presented but infinite options could be implemented.

Finally, there are many scopes for improvement, such as modify the available actions according to the vehicle state so it can choose the feasible actions or it also would be interesting to try to train the model in the experimental scenarios and include some physical constrains that are not considered or neglected in the vehicle's behavior. Also, a longitudinal controller can be implemented to create a speed reference. Some improvements of the low-level controllers can be made to have straighter paths on the track and also while training the model it would be better to include large heading errors for sharp curvatures and replicate this for other tracks.

References

1. Emuna R, Borowsky A, Biess A (2020) Deep reinforcement learning for human-like driving policies in collision avoidance tasks of self-driving cars. arxiv:2006.04218
2. Kiran BR, Sobh I, Talpaert V, Mannion P, Sallab S, Yogamani K, Pérez P (2020) Deep reinforcement learning for autonomous driving: a survey. arxiv:2002.00444
3. Li D, Zhao D, Zhang Q, Chen Y (2018) Reinforcement learning and deep learning based lateral control for autonomous driving. arxiv:1810.12778
4. Matute J, Marcano M, Diaz S, Pérez J (2019) Experimental validation of a kinematic bicycle model predictive control with lateral acceleration consideration. 52:07. https://doi.org/10.1016/j.ifacol.2019.08.085
5. Naveed KB, Qiao Z, Dolan JM (2020) Trajectory planning for autonomous vehicles using hierarchical reinforcement learning. arxiv:2011.04752
6. Polack P, Altché F, d'Andréa Novel B, de La Fortelle A (2017) The kinematic bicycle model: a consistent model for planning feasible trajectories for autonomous vehicles. In: 2017 IEEE intelligent vehicles symposium (IV), pp 812–818. https://doi.org/10.1109/IVS.2017.7995816
7. Puterman ML (2014) Markov decision processes: discrete stochastic dynamic programming. Wiley
8. Rajamani R (2006) Vehicle dynamics and control. ISBN 0-387-26396-9. https://doi.org/10.1007/0-387-28823-6
9. Rojas MG, Olivera AC, Vidal PJ (2022) Optimising multilayer perceptron weights and biases through a cellular genetic algorithm for medical data classification. Array 14:100173. ISSN 2590-0056. https://doi.org/10.1016/j.array.2022.100173. URL https://www.sciencedirect.com/science/article/pii/S2590005622000339
10. Singh S (2015) Critical reasons for crashes investigated in the national motor vehicle crash causation survey
11. Stang M, Grimm D, Gaiser M, Sax E (2020) Evaluation of deep reinforcement learning algorithms for autonomous driving. In: 2020 IEEE intelligent vehicles symposium (IV), pp 1576–1582. https://doi.org/10.1109/IV47402.2020.9304792

12. Sutton RS, Barto AG (2018) Reinforcement learning: an introduction. a bradford book. Cambridge, MA, USA, p 0262039249
13. Szoke L, Aradi S, Becsi T, Gaspar P (2020) Vehicle control in highway traffic by using reinforcement learning and microscopic traffic simulation. In: 2020 IEEE 18th international symposium on intelligent systems and informatics (SISY), pp 21–26. https://doi.org/10.1109/SISY50555.2020.9217076
14. Vu T-D (2009) Vehicle perception: localization, mapping with detection, classification and tracking of moving objects. Theses, Institut National Polytechnique de Grenoble—INPG. URL https://tel.archives-ouvertes.fr/tel-00454238
15. Watkins C, Dayan P (1992) Technical note: Q-learning. Mach Learn 8:279–292. https://doi.org/10.1007/BF00992698
16. Wiering M, van Otterlo M (2014) Reinforcement learning: state-of-the-art. Springer Publishing Company, Incorporated, p 364244685X
17. William F, Milliken D-LM (1995) Race car vehicle dynamics. Society of Automotive Engineers. Warrendale, Pa

Performance Evaluation of Machine Learning App Approach to Modular Arrangement of Predetermined Time Standard

Emmanuel Basitere, Ilesanmi Daniyan, Khumbulani Mpofu, and Adefemi Adeodu

Abstract Merging Modular Arrangement of Predetermined Time Standard (MODAPTS) and techniques used in the fourth industrial revolution (4IR) such as Machine Learning (ML) can start to improve the user experience of time standards. This study used Artificial Neural Networks (ANN) applied as chatbots to see whether ML could indeed improve MODAPTS in terms of ease, pace, and accessibility. The conventional and ANN methods were compared with assistance from logistics engineers, and the ANN approach. A chatbot using ANN was created and packaged on an html page presented to the research participants making use of a mobile device. The experiment used five material handling written scenarios to emulate the observation process, looking at the traditional approach when conducting a MODAPTS time study then followed by the ANN solution making use of the chatbot. ANN was found to be 0.25 min faster at a prediction rate of over 90% when the chatbot was in use. The result showed that machine learning could indeed be used with MODAPTS to equal performance and potentially improve the use of the time standard. The neural network was able to accurately predict the MODAPTS code of 94.7% of the 262 activities entered by the research participants. The potential to add other ML learning techniques and time study methods exists, the template is flexible enough to be moulded into a tool that all engineers can adapt in their different working environment.

Keywords ANN · Chatbot · 4IR · MODAPTS

E. Basitere · I. Daniyan (✉) · K. Mpofu
Department of Industrial Engineering, Tshwane University of Technology, Pretoria, South Africa
e-mail: daniyan.ia@achievers.edu.ng

A. Adeodu
Department of Mechanical Engineering, University of South Africa, Florida, South Africa

G. L. Conte and O. Sename (eds.), *Proceedings of the 11th International Conference on Mechatronics and Control Engineering*, Lecture Notes in Mechanical Engineering,
https://doi.org/10.1007/978-981-99-6523-6_9

1 Introduction

Conducting a MODAPTS analysis usually involves several components that would allow for a successful feasibility study, this includes but not limited to understanding an activity by video analysis or continuous observation, analysing possible efficiencies from that observation, and most importantly an in-depth knowledge of the MODAPTS time standard and its principles to prove those efficiencies. Mastering the MODAPTS time standard can be a bit challenging and rather complicated for the engineer [1]. The total time it takes to conduct a MODAPTS study may take too long due the nature of the observed activity and converting those activity tasks to MODAPTS code. MODAPTS is rather an old-fashioned time standard when looking at its inception and application, but efforts were made to simplify the time standard in multiple environments. Applications such as clerical, transit, janitorial and sewing are some of the options an engineer may explore. These applications are dated and too wordy, specifically when looking to solve a problem as an engineer today, in this era of 4IR. Suffice to say the world has changed drastically today from the 1960's era to the era of the 4IR. Furthermore, the adoption of Artificial Intelligence (AI) has made it easier to solve difficult problems experienced in the past [2–4].

Merging the MODAPTS time standard with solutions birthed from 4IR such as machine learning, can help reduce the complexity and improve efficiency when conducting feasibility studies. A notable example is research conducted by Wu et al. [5], who proposed incorporating motion analysis to the MODAPTS time standard by adding an extensive ergonomics template called Principle Components Analysis (PCA). The study found that the PCA-based approach was rated 80.08% amongst its participants and was more efficient at three minutes as opposed to the one hour that it took using the traditional MODAPTS approach. Another notable inclusion was a wearable sensory glove proposed by Mallembakam [1], the research focused on using a Bluetooth module mounted onto a glove that translated hand movements into MODAPTS code, this was compared to the traditional MODAPTS approach and it was found that the glove showed reliable and reputable results with that of using MODAPTS the conventional way.

An untapped field that would be an impactful inclusion is the merger of MODAPTS and machine learning. Today machine learning is more prevalent making it easier to solve difficult problems and improve livelihoods [6–8]. A massive advantage to researching machine learning techniques is that they are largely open sourced when looking at big data [9], and can be applied to the MODAPTS time standard.

The significance of this study is that a consolidation of multiple MODAPTS process keywords was collected for continual neural network training. Furthermore, a potential pattern recondition system to assist the engineer during observations may emerge through continues use of the ANN research template. Hence, the research looks at a form of Machine Learning called Artificial Neural Networks (ANN) as an application to MODAPTS with the aim to improve the activities involved during time study observations whilst making use of the MODAPTS time standard. Hence, this research will look to solve the issue of rapidity and the ease of use in conducting a MODAPTS study making use of machine learning.

2 Methodology

A chatbot using ANN was created and packaged on an html page presented to the research participants making use of a mobile device. The experiment used five material handling written scenarios to emulate the observation process, looking at the traditional approach when conducting a MODAPTS time study then followed by the ANN solution making use of the chatbot.

The chatbot measured each research participant against several key attributes that defined the user experience toward the chatbot application. The experiment used a simple NN chatbot to employ a MODAPTS conversation with the research participants. The focus was on the predicted MODAPTS code produced by the chatbot, each time a research participant presented an activity. Primary data was collected from MODAPTS time studies produced by DSV and was used to train the chatbot. This was mostly dialect in the form of keywords showing how DSV described their process activities.

The chatbot was build making use of the python programming language, it used a ML module called Keras to train and implement the NN. The python code was executed using an integrated development environment called PyCharm.

2.1 The Chatbot Intent File

Constructing a chatbot consisted of several programming files working together to create the application. The chatbot was first initialised with a JavaScript Object Notation file called "intents.js". This file defined the possible intentions that could take place during interactions with the chatbot [9–13]. Figure 1 shows a snippet of the "intents.js" file.

The intents file consisted of a tag element i.e., a subcategory that a user query might fall into. Each terminal class "get" and "put" were paired with a movement class element and the degree of difficulty to form the tag. Any query for example that used the "hand" movement to "get" (terminal) an object with an "easy" degree of difficulty would fall under the "hand_get_easy" tag or subcategory.

The next element in the intens.js file was the patterns, a group of keywords that a user query must contain to fall under a specific tag or subcategory. The patterns keywords were created using synonyms of the terminal class (get, put) paired up with the movement class and the degree of difficulty. A user query "Pick box using your hand at an easy stance" will fall under the "hand_get_easy" category as the keywords "pick"," hand" and "easy" are present in the query.

```
{
  "tag": "hand_get_easy",
  "patterns": [
    "obtain hand easy",
    "collect hand easy",
    "get hand easy",
    "take hand easy",
    "acquire hand easy",
    "pick hand easy",
    "hold hand easy",
    "gain hand easy",
    "attain hand easy"
  ],
  "responses": [
    "M2G1"
  ]
},
```

Fig. 1 Code snippet of the "intents.js" file

Once a user query falls under a tag, then the chatbot will issue a response found under the responses keyword. The response of the user query was the MODAPTS code designated for the tag [14].

Table 1 shows an example of a general query made to the chatbot NN.

The chatbot was able to break down sentences if the user was able to split activities of the sentence using a period, comma or the "And" keyword. Each research participant was given this example including a list of rules and conversion table to use when querying the chatbot.

Table 2 shows the MODAPTS symbol conversion when querying the Chatbot NN.

Table 1 Example of a chatbot query

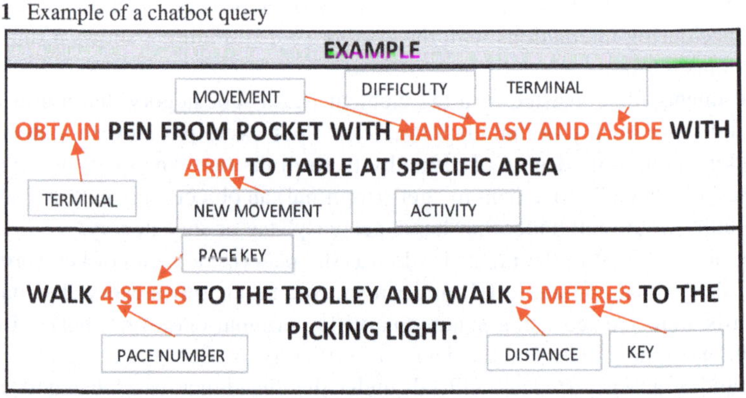

Table 2 Conversion table

	SYMBOL	DESCRIPTION
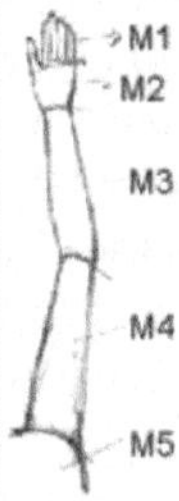	**M1**	**Fingers**
	M2	**Hand**
	M3	**Arm**
	M4	**Shoulder**
	M5	**Extended Shoulder**
	M7	**Trunk**
DIFFICULTY	**G1**	**Easy**
	G3	**Difficult**
	P0	**General**
	P2	**Specific**
	P5	**Exact**
	**W5**	**Steps**
	W7.75	**Metres**

2.2 *Deployment of the Experiment*

The chatbot was deployed to the research participants making use of a webpage using a mobile device. The code for the chatbot was packaged and stored on a server and accessed on request by the web page using the flask micro web framework. A user request would be captured on webpage, then sent to the server to query the chatbot NN, then perform some MODAPTS calculations and then finally returns the results to the user on the webpage.

The front-end used a styling toolkit called Bootstrap to position and place the elements onto the webpage. The backend contained a database to hold user actions, MODAPTS data and user information while the application was being used.

Each research participant was first given a questionnaire asking basic experiential questions relating to MODAPTS. Once the questionnaire had been completed then the participants were given the five scenarios to read and conduct a MODAPTS study with the acquired information. A stopwatch was used to measure the total duration that each research participant took to complete the MODAPTS study for each scenario and was recorded.

The next step involved giving each participant an android device with the machine learning webpage. They each were required to sign up to use the application which was helpful as their tracked user data could be stored and accessed anytime and anywhere with aid of a local database. They each assessed the MODAPTS scenarios again this time making use of the ML App. The scenarios were built into the application and were accessed through a touch of a button. The application was able to record MODAPTS durations from the time each research participant logged onto the application until the time they had signed off.

The data generated from the application by the participants were collected and simultaneously sent to a database preparing them for analytical studies.

Acquiring this information required building an ANN chatbot to interact with the user, developing a web interface for that chatbot and finally, employing elements in the application that would keep record of the statistics needed to define that potential improvement.

There were setbacks in acquiring some primary data but ultimately scenarios were created to facilitate that shortfall. The prerequisites of the experiment took time to compile but was necessary to extract crucial information about the research participants. All these elements individually contributed to finding the total time to compare MODAPTS and the ML approach.

2.3 *Testing the Chatbot Neural Network*

Testing the chatbot involved creating a python file that would access the saved model, submit the user query, and give the prediction. Each prediction was assessed based on the intents file, if an error occurred then the intents file was adjusted, and the model was re-trained. This repeated action was done on the terminal of the IDE and then applied to the UI once the testing was completed and chatbot acceptable for use. Figure 2 shows the testing code snippet.

3 Results and Discussion

The section is divided into two parts, first part focused on the results and analysis of the Machine Learning App approach of conducting the same MODAPTS study while the last part compared the two approaches in relation to the final duration with aid of a t-test analysis.

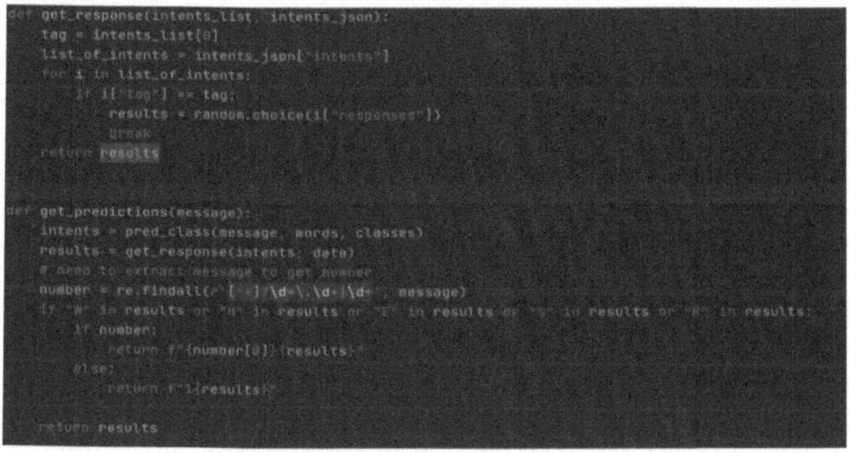

```
def get_response(intents_list, intents_json):
    tag = intents_list[0]
    list_of_intents = intents_json["intents"]
    for i in list_of_intents:
        if i["tag"] == tag:
            results = random.choice(i["responses"])
            break
    return results


def get_predictions(message):
    intents = pred_class(message, words, classes)
    results = get_response(intents, data)
    # need to extract message to get number
    number = re.findall(r'[-+]?\d+\.\d+|\d+', message)
    if "[illegible]" in results or "[illegible]" in results or "[illegible]" in results or "[illegible]" in results or "[illegible]" in results:
        if number:
            return f"{number[0]}{results}"
        else:
            return f"1{results}"

    return results
```

Fig. 2 Code snippet for testing the NN chatbot

3.1 Error Accumulation

Error accumulation monitors all errors generated by the research participants while making use of the application. The frequency of each error correlates to a difficulty in use of the application or a particular deficiency in understanding the MODAPTS language from the research participant. Errors monitored included: editing an activity, deleting an activity, input length error and MODAPTS code error.

Table 3 depicts the error entries made by each research participant throughout all scenarios.

There was a total of thirty-two errors done by the research participants. None of these errors contributed significantly to the functionality of the application but impacted the total time. The error for "description length" was encountered the most, a total of nineteen times appearing in four of the five scenarios. The error was a timestamp recorded each time the prediction button was pressed with no activity added. This mistake was expected as the prediction button needed to be pressed to activate the chatbot. Figure 3 shows a clustered column graph of the accumulated application errors.

Table 3 Error entities by scenario

Activity	SC1	SC2	SC3	SC4	SC5	TTL
Code			3		1	**4**
Delete			2	1	2	**5**
Description length	5	7	2	5		**19**
Activity		2	1	1		**4**
Total	**5**	**9**	**8**	**7**	**3**	**32**

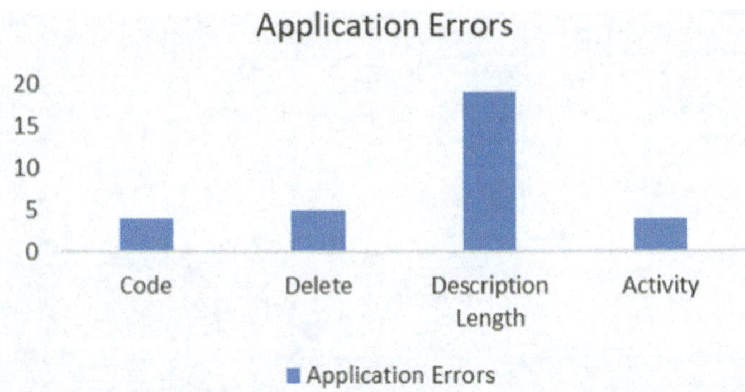

Fig. 3 Application errors

3.2 Total Duration

The total duration is a measure that analyses the total time it takes to conduct a MODAPTS study using the NN application. A timestamp is added each time the research participant log's in or sign's up into the NN application. Another timestamp is added each time a research participant logs out of the application. The two timestamps are the subtracted from each other to gain the total duration through by scenario.

Table 4 shows a snippet of the entries entered each time a user log's in, sign's in and log's out of the application.

The total average time each research participant took to perform a study using the chatbot NN was 11.2 min. Scenario 5 displayed longest time, but this was caused by an outlier from research participant 4 with the higher duration of 21 min. The lowest recorded total average and duration was scenario 1, this was expected as the scenario had the least number of activities, and its aim was to gently introduce the participants to the chatbot application.

Table 4 Activity durations

Participant	SC1	SC2	SC3	SC4	SC5	TTA	TTD
1	6	9	10	19	11	11	55
2	6	9	7	13	10	9	45
3	7	9	6	6	11	7,8	39
4	9	22	20	13	21	17	85
Total AVR	**7**	**12,25**	**10,75**	**12,75**	**13,25**	**11,2**	
Total duration	**28**	**49**	**43**	**51**	**53**		**224**

3.3 Discussion of the Traditional and ML Approach Results

The research participants reacted well to conducting a MODAPTS study under traditional means, this was expected as it is the norm. Introducing the ML application was at first a learning curve but gradually became easier through each scenario as shown by a 41.2% decrease in App navigation time from each participant. Hence, the notion mentioned by Wu et al. [5] with regards to a learning deficiency that new engineers face when presented with MODAPTS becomes less challenging and more manageable during study creation. A novice engineer can focus on describing the activity rather than the complex steps involved in MODAPTS code formulation.

The neural network was able to accurately predict the MODAPTS code of 94.7% of the 262 activities entered by the research participants. This is in line with the findings of existing studies that have indicated the suitability of the neural network for predictive purpose [15–17] This aspect saved a considerable amount of time that the research participants would have used to formulate the MODAPTS code. The total average time each research participant took to perform a study using the chatbot NN was 11.2 min. Furthermore, when looking at phase 3 of the framework developed by Kumar et al. [18], activity formulation becomes simpler due to the library of keywords that initially and consistently get added through the continuous use of the ANN.

The App alternative proved to be 0.25 min faster than the traditional approach when used under the same conditions. The difference may be small, but the ML approach performed well to produce equal results under first time users as compared to the already establish excel template used by an adept participant. This was echoed by the approach followed by Mallembakam [1] and Wu et al. [5] who implemented innovative tools along with the MODAPTS time standard to match and even reduce the duration of conducting MODAPTS time studies. This proves that MODAPTS can be paired external solutions to better the performance of the time standard.

The ML chatbot however had challenges understanding some activities as it had a deficiency when deciphering a particular group of MODAPTS code. The group consisted of the element's "R", "E" and the finger movement, which caused the application to record an error or forced the research participant to manually input the code forfeiting the prediction.

Furthermore, the application had no method of handling repetition when inputting an activity. This drastically increased the time that each research participant spent when using the ML alternative during the experiment. All code related errors have been fixed and the potential of decreasing the final ANN time is evident when the consistent repetition is removed from the application.

4 Conclusion

The objective of this study was to test the neural network using operational scenarios against the traditional MODAPTS observation approach. This was achieved by packaging the created ANN chatbot template into an html page and handing that to the research participants to use and test. Five written scenarios were also used to traditionally create the MODAPTS time study using four activities. These two approaches were compared with each other to find the most effective and fastest method.

Furthermore a user-friendly and workable ANN MODAPTS template was developed. This was achieved as the chatbot was packaged onto an HTML page and deployed onto a mobile device. The template was built to be user friendly geared for the research participant as it made use of a styling framework called bootstrap. The results obtained indicated that the ANN was found to be 0.25 min faster at a prediction rate of over 90% when the chatbot was in use. The result also showed that machine learning could indeed be used with MODAPTS to equal performance and potentially improve the use of the time standard. The neural network was able to accurately predict the MODAPTS code of 94.7% of the 262 activities entered by the research participants. This saved a considerable amount of time that the research participants would have used to formulate the MODAPTS code. The total average time each research participant took to perform a study using the chatbot NN was 11.2 min. The activity formulation becomes simpler due to the library of keywords that initially and consistently get added through the continuous use of the ANN.

The time difference may be small, but the ML approach performed well to produce equal results under first time users as compared to the already establish excel template used by an adept participant.

The ANN approach makes it far easier to implement MODAPTS in multiple industries due to the chatbots customisable keywords and dialect recognition. A wealth of MODAPTS data is far more easily reachable as the template would be online to be used. The results show that MODAPTS ML template replaces tedious MODAPTS analysis done on the computer with a mobile alternative that consistently records user and study analytics for the betterment of overall analysis. The template has great potential for improvement and refinement when looking at efficiency, it makes data sharing amongst engineers easy due to its descriptive database.

The potential to add other ML learning techniques and time study methods exists, the template is flexible enough to be moulded into a tool that all engineers can adapt in their different working environment not just only automotive. Future works can test the neural network against the traditional MODAPTS observation approach using other operational scenarios for more performance evaluation.

Acknowledgements Funding: "The authors disclosed receipt of the following financial support for the research: Technology Innovation Agency (TIA) South Africa, Gibela Rail Transport Consortium (GRTC), National Research Foundation (NRF grant 123575) and the Tshwane University of Technology (TUT)."

References

1. Mallembakam VR (2020) Incorporating modular arrangement of predetermined time standard with a wearable sensing glove. University of Windsor, Canada
2. Aissani N, Beldjilali B, Trentesaux D (2008) Use of machine learning for continuous improvement of the real time heterarchical manufacturing control system performances. Int J Ind Syst Eng 3:474–497
3. Alpaydin E (2020) Introduction to machine learning. MIT Press
4. Brownlee J (2017) Deep learning for natural language processing: develop deep learning models for your natural language problems. Machine Learning Mastery
5. Wu S, Wang Y, Bolabola JZ, Qin H, Ding W, Wen W, Niu J (2016) Incorporating motion analysis technology into modular arrangement of predetermined time standard (MODAPTS). Int J Ind Ergon 53:291–298
6. Donkers T, Loepp B, Ziegler J (2017) Sequential user-based recurrent neural network recommendations. In: Proceedings of the eleventh ACM conference on recommender systems, pp 152–160
7. Manaswi NK (2018) Developing chatbots. In: Deep learning with applications using Python. Springer
8. Zaremba W, Sutskever I, Vinyals O (2014) Recurrent neural network regularization. arXiv preprint arXiv:1409.2329
9. Landset S, Khoshgoftaar TM, Richter AN, Hasanin T (2015) A survey of open source tools for machine learning with big data in the Hadoop ecosystem. J Big Data 2:24
10. Brandtzaeg PB, Følstad A (2017) Why people use chatbots. In: International conference on internet science. Springer, pp 377–392
11. Lishchynska D (2017) What are bots, how do chat bots work? What are bots, how do chat bots work? [Online]. Available from: https://botscrew.com/blog/what-are-bots/#:~:text=Chatbot%20or%20bot%20%E2%80%93%20is%20a,an%20instant%20pre%2Dset%20answer. [Accessed 02/01 2021]
12. Nelson D (2021) Text generation with Python and Tensor/Flow. Available from: https://stackabuse.com/text-generation-with-python-and-tensorflow-keras [Accessed 07/30 2021]
13. Pykes K (2021) Build a simple chatbot in Python with deep learning. Available from: https://towardsdatascience.com/a-simple-chatbot-in-python-with-deep-learning-3e8669997758 [Accessed 31/03/2021 2022]
14. Stewart JR (2002) Applying MODAPTS standards. Soc Work Sci (SWS) 1:1–4
15. Adesina OT, Jamiru T, Daniyan IA, Sadiku ER, OgunbiyI OF, Adesina OS, Beneke LW (2020) Mechanical property prediction of SPS processed GNP/PLA polymer nanocomposite using artificial neural network. Cogent Eng 7(1720894):1–17
16. Daniyan IA, Mpofu K, Tlhabadira I, Ramatsetse BI (2021) Process design for milling operation of titanium alloy (Ti_6Al_4V) using artificial neural network. Int J Mech Eng Rob Res 10(11):601–611
17. Daniyan IA, Bello EI, Ogedengbe TI, Mpofu K (2020) Use of central composite design and artificial neural network for predicting the yield of biodiesel. Procedia CIRP 89:59–67
18. Kumar R, Charak A, Thakur G (2020) Productivity improvement of an automotive assembly line using modular arrangement of predetermined time standards (MODAPTS). i-Manager's J Future Eng Technol, 16, 32

Waveform Tracker Alarm for Automatic Patient-Ventilator Asynchrony (PVA) and Mechanical State Recognition for Mechanical Ventilators Using Embedded Deep Learning

Paul Ryan A. Santiago, Paul M. Cabacungan, Carlos M. Oppus, John Paul A. Mamaradlo, Neil Angelo M. Mercado, Reymond P. Cao, and Gregory L. Tangonan

Abstract The Ateneo Innovation Center designs and develops a modular approach to medical alarm and alert systems for mechanical ventilators that enable clinicians to remotely monitor patient conditions and ventilator circuit status in near real-time, providing decision support that allows for a better diagnosis. It monitors and tracks the alarm events related to the ventilator waveform consisting of pressure, flow, and volume curves by using automatic peak detection of the curves and real-time recognition of time-series waveforms. The developed system combines the threshold alarms with embedded Artificial Intelligence to automatically detect complex alarms that need medical expertise such as issue detection on asynchrony, anomalies, and mechanical. It also differentiates the critical types of alarms, assisting clinicians via alarm prioritization, and remote patient monitoring via a near cloud system. Storing data in the near cloud system as a medical database enables building a rich dataset for upgrading the predictive model of alarm recognition.

Keywords Emergency ventilator · Biomedical monitoring · TinyML · Near cloud · And circular electronics manufacturing

P. R. A. Santiago (✉) · C. M. Oppus · G. L. Tangonan
Ateneo de Manila University, 1108 Quezon City, Philippines
e-mail: paul.santiago@obf.ateneo.edu

P. M. Cabacungan · C. M. Oppus · J. P. A. Mamaradlo · N. A. M. Mercado · R. P. Cao · G. L. Tangonan
Ateneo Innovation Center, 1108 Quezon City, Philippines

G. L. Conte and O. Sename (eds.), *Proceedings of the 11th International Conference on Mechatronics and Control Engineering*, Lecture Notes in Mechanical Engineering,
https://doi.org/10.1007/978-981-99-6523-6_10

1 Introduction

In 2020, the COVID-19 pandemic disrupted the world by spreading at unprecedented rates and causing tens of thousands of fatalities within a few months [1]. Even with vaccines, its mutations unpredictably develop into various strains, and the number of infections and deaths are still on the rise, especially in regions where the number of patients in need of hospital care exceeds the availability of care. According to the Office of Inspector General for the U.S. Department of Health and Human Services, hospitals have reported a scarcity of skilled physicians needed to meet the anticipated patient surge. Many hospitals also stated that they lacked trained personnel who could operate ventilators and treat patients requiring that degree of care [2].

A mechanical ventilator machine is a life-support device, when the machines record measurements outside of normal parameters, it beeps, and alarms ring out to alert medical staff to potential problems. The data from the bedside monitor is usually lost as the monitor screen refreshes every few seconds. It requires intensive monitoring to identify early signs of clinical worsening and to minimize the risk of iatrogenic harm [3, 4]. With A-vent [5], the efforts of the Ateneo Innovation Center (AIC) to design, develop and operate a modular and low-cost ventilator alarm were described. The updated system currently triggers an alarm with the patient-ventilator asynchrony (PVA), anomalies, and mechanical problems as the previous system's alert system was limited only to its waveform parameters such as pressure, flow, and volume that alerts clinicians when the parameters fall below or above the set limits. This development of an alarm system is a design and engineering study with no humans involved.

2 Review of Related Literature

"Fighting the ventilator" is a common occurrence when the patient's demand does not match the machine's delivery, one of the reasons users' training is necessary to assure positive patient outcomes [3, 6]. The interaction between the patient and the machine is difficult to manage, hence the ventilator should be synced with the patient's normal inhalation and exhalation cycles.

Different ventilator designs emerged worldwide during this time of the pandemic. Corey et al. [7] presented a low-cost and easy-to-produce electronic sensor and alarm system for pressure-cycled ventilators that utilized an algorithm inspired by those used in hearing aids that required little memory that it can run on a microcontroller. The device estimated clinically useful metrics such as pressure and respiratory rate and sounds an alarm when the ventilator malfunctions. The application of the Internet of Things (IoT) protocol on medical equipment, as demonstrated by Mashoedah et al. [8], was intended to protect medical workers dealing with COVID-19 patients, particularly while medical personnel is monitoring and setting up such devices. Data was collected through testing, observation, and limited field tests using their "Define, Design, Develop, and Disseminate (4D)" approach. Rehm et al. [9] developed an intelligent decision support system using a Raspberry Pi that collects data from the ventilator unit and was able to store the stream of ventilator waveform and physiological data and analyzed it using supervised Machine Learning (ML) to classify the double triggering, breath stacking asynchronies, and acute respiratory distress syndrome (ARDS). It used IoT wireless connectivity to visualize the ventilator waveform and relied on a cloud platform to store and process the data.

The current ventilator system does not have a self-monitoring feature, which is critical for ensuring that the ventilator machines are working properly and that the settings are appropriate for the patient's conditions in real-time. This prompted the team to spearhead and start this project.

3 System Description

3.1 Experimental Setup

The conventional ventilator machines are threshold-based alarms that are prone to frequent false alerts. There are currently no intelligent systems embedded in emergency ventilators to automatically detect cycling asynchrony and generate alerts to clinicians. This study presented a new approach in which the patient and ventilator interactions characterized by a stream of ventilator waveform data were recognized in a real-time and stand-alone manner. Figure 1a describes the simple design of the A-vent unit and its experimental setup for emulating the different alarm events, including types of PVA, ventilator airway circuit status, anomalies, and high/low threshold levels occurrence.

The supplied air goes into the patient's airway circuit through a 1 L test lung that mimics a patient's lungs. The experimental setup is subject to a constant air supply and a one-second inspiratory and expiratory ratio. When the ventilator unit delivers pressurized air, the test lung expands and contracts accordingly with the given inspiratory and expiratory (I:E) ratio. The alert events are emulated as a proof of concept. The methods for emulating the patient and ventilator interactions are given in Table 1. Asynchrony and ventilator circuit-related alerts are the two types of modeled patient and ventilator interaction alarms. The emulated waveforms consist of eight classes, labeled as (1) normal waveform (NW), the common types of PVA include (2) delay cycling (DC^1), (3) double triggering (DT^1), (4) reverse triggering (RT^1), and (5) ineffective effort (IE^1) and alerts related to ventilator airway circuit status include (6) disconnected pressure port (DPP^2), (7) disconnected tube (DT^2), and (8) machine failure (MF^2).

To create a unique pattern of waveforms, the patient and ventilator interactions are modeled by altering the open-close state of the ventilator unit's manual air release valve during the inspiratory period, disconnecting components of the airway circuit, and shutting down the ventilator unit. The emulated asynchrony waveforms are

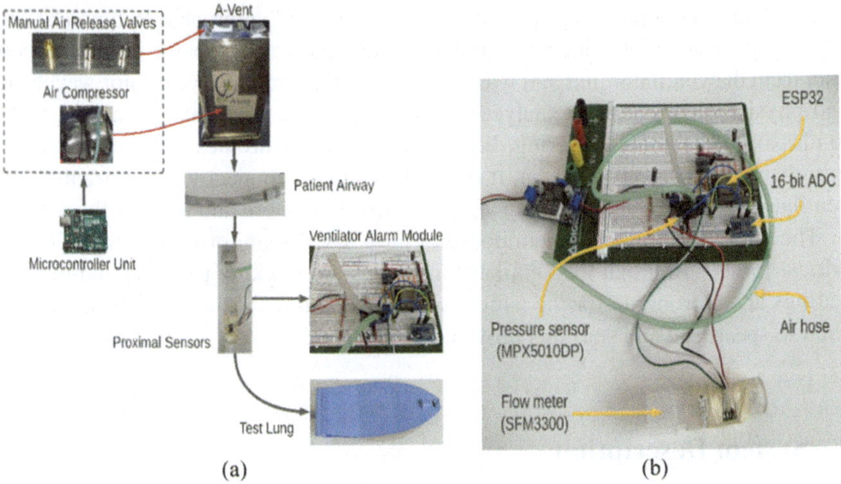

Fig. 1 Experimental setup **a** for emulating and capturing ventilator waveform data, and **b** modular intelligent ventilator alarm system prototype

Table 1 Summary of emulated ventilator waveform

Labels	Types of asynchronies	Emulation method
NW	Normal waveform	Valve is opened a little bit
DC[1]	Delay cycling	Valve is opened then close
DT[1]	Double triggering	Valve is closed, opened, then closed
RT[1]	Reverse triggering	Valve is closed then opened
IE[1]	Inefficient effort	Valve is fully opened
DPP[2]	Disconnected pressure port	The pressure sensor is disconnected
DT[2]	Disconnected tube	Test lung from the tube is removed
MF[2]	Machine failure	The ventilator unit is shut down

[1] Alarms related to common types of PVA
[2] Alarms related to ventilator airway circuit status

comparable to actual types of PVA associated with cycling and patient effort criteria and have been evaluated by a physician. The rest are machine and airway circuit issues such as power failure, air hoses, and circuit tube disconnection.

3.2 *Alarm Algorithm*

We achieved significant improvements to a conventional ventilator alarm system in this study by embedding Artificial Intelligence (AI) within a sensor-equipped

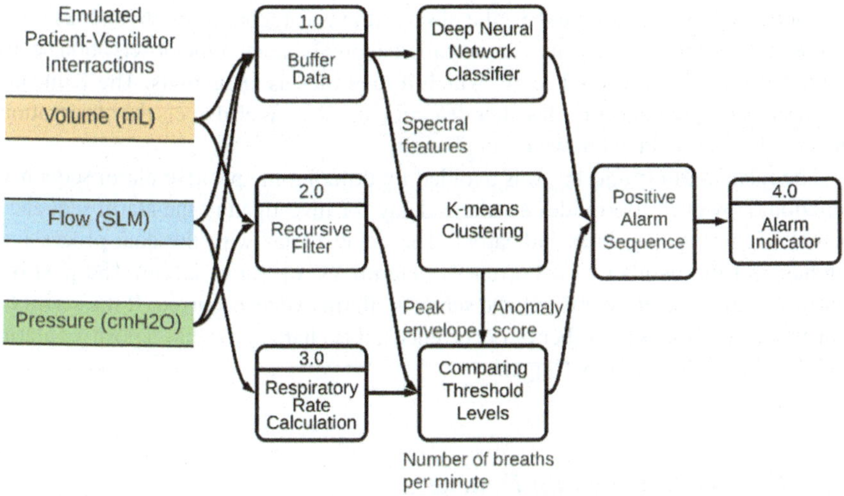

Fig. 2 Data flow for A-vent modular intelligent alarm device

ventilator machine. We performed built-in testing of the ventilator operation and analyzed its waveform for the recognition of time-series alarm events. When deviations from regular operations occur, ML together with data processing and sequencing algorithms alert medical staff.

Figure 2 illustrates the data flow describing the processing and algorithms of the alarm module for detecting the critical and important types of alerts. The ventilator waveforms comprise three parameters which include (1) pressure, (2) flow and (3) volume that was captured by a medical-grade flow meter and pressure sensor. The real-time data from the sensors are processed by the microcontroller unit. There are three algorithms used to develop the intelligent ventilator alarm module: (1) a couple of recursive filter algorithms, (2) K-means clustering, and (3) a deep neural network. Each algorithm specializes in detecting different types of alarms.

We employed the study of Corey et al. [8] to track the peak-to-peak pressure cycling (i.e., PIP, and PEEP) and the peak tidal volume. The respiratory rate is calculated by measuring the inspiratory period with the number of breaths per minute given by the I:E ratio. The PIP, PEEP, peak tidal volume, respiratory rate, and anomaly score are the five parameters for the threshold-based alarm. The decisions of these alarms are based on whether the current parameter falls above or below the set thresholds.

On the other hand, the raw data needed to be buffered for the processing which converts the time-series data to data suitable for ML algorithms. The spectral analysis used the extracted features of the raw data to model the PVA and mechanical state and then feed it to the deep neural network classifier. The basis of the decision for alarms was the predictions of the model represented by labels and accuracy.

The K-means clustering algorithm was used to find the natural pattern of the data and to detect anomaly data from the dataset. If the ventilator waveform data samples do not belong to any data clusters, the observation is categorized as anomalous [10]. The K-means anomaly returns a value called anomaly score if the observation score is greater than the threshold score, which it identifies as anomalous. The K-means clustering complements the classifier detection model, which detects the observation outside the dataset also known as an anomaly.

The false alarm triggering was avoided by utilizing the positive alarm sequence function. However, it provides an alarm delay for investigating the alarm sequence before triggering the alarm indicators. The delay varies with the data processing latency and the number of occurrences determines a positive alarm. The positive alarm sequence function ensures the series of alarm events occurred. If the series of alarms exceeds the set number of occurrences, it is characterized as a positive alarm and the alert indicator may trigger.

3.3 Data Gathering and Dataset

The ventilator waveform is captured using a medical-grade Sensirion flow meter (SFM3300) and differential pressure sensor (MPX5010DP) are shown in Fig. 1b. Figure 3 shows the ventilator waveforms comprising three parameters which include (1) pressure, (2) flow and (3) volume, which are captured by the sensors interfaced to a microcontroller.

The MPX5010DP is a differential pressure sensor designed to interface with a microcontroller or microprocessor that has an analog to digital (A/D) converter. It is an analog device with a high-resolution analog voltage signal ranging from 0 to 5 V that are proportional to the applied pressure of 0 to 10 kPa. The pressure is proportional to the output voltage, the measured pressure $P_{\mathrm{cmH_2O}}$ in centimeter of water (cm-H_2O) can be described as:

$$P_{cmH_2O} = \left(\left(V_{out} - V_{offset}\right)\right)/Sensitivity/10 \tag{1}$$

where the V_{out} is the output voltage of the pressure sensor in millivolts which is fed to a 16-bit A/D converter. The parameters offset voltage, V_{offset}, and *Sensitivity* which values can be seen in the operating characteristic section in the datasheet is 0.2V and 4.413 mV/mmH_2O respectively. The Sensirion SFM3300 is a digital and bidirectional flow sensor for proximal flow measurement in respiratory applications that can measure a flow range of ±250 standard liters per minute (SLM). Based on the product technical specification the flow F_{SLM} measured in SLM is described as:

$$F_{SLM} = \left(value_{I2C} - value_{offset}\right)/scale\ factor \tag{2}$$

where the $value_{I2C}$ is the integer return value by the flow meter from the I2C communication interface. The parameter $value_{offset}$ and *scale factor* (1/SLM) can be seen in the electrical characteristic section of the product specification, where the given values are 32,768, and 120 respectively. The calculation of tidal volume was derived

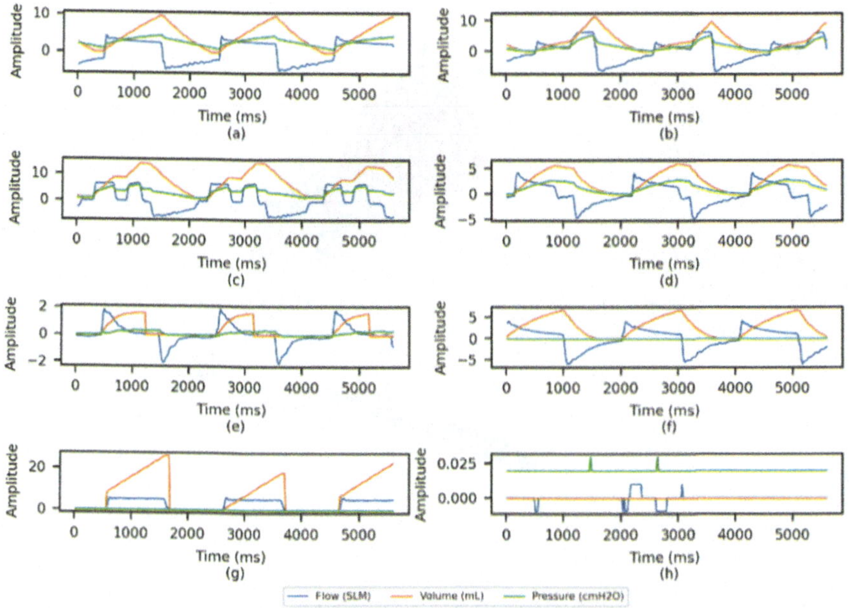

Fig. 3 The pressure, flow, and volume emulated ventilator waveforms as captured by the sensors; **a** normal waveform, **b** delay cycling, **c** double triggering, **d** reverse triggering, **e** inefficient effort, **f** disconnected pressure port, **g** disconnected tube, and **h** machine failure

from flow measurements. Given that the A-vent unit delivers pressurized air at the rate of a one-second I:E ratio, the tidal volume TV_{mL} measured in millimeters (mL) is described as:

$$TV_{ML} = \sum F_{SLM}/60 * \Delta t \quad (3)$$

where the F_{SLM} is the flow rate expressed in standard liters per minute, and Δt is the sampling interval obtained from the sampling rate. The continuous time-series waveform data are stored as comma-separated values (CSV) files with the timestamp in millisecond intervals given by the sampling rate to create a dataset. The emulated waveform was sampled at the rate of 50 Hz and captured continuously for 10 min for each class. The dataset was randomly divided into training, validation, and testing set. Before training the PVA and machine state recognition model, the dataset was processed to reduce its samples represented by its features as inputs to ML algorithms.

3.4 Features Extraction

Embedded devices such as microcontrollers have limited computational power and memory, making it vital to optimize the processing of large amounts of data. Feature

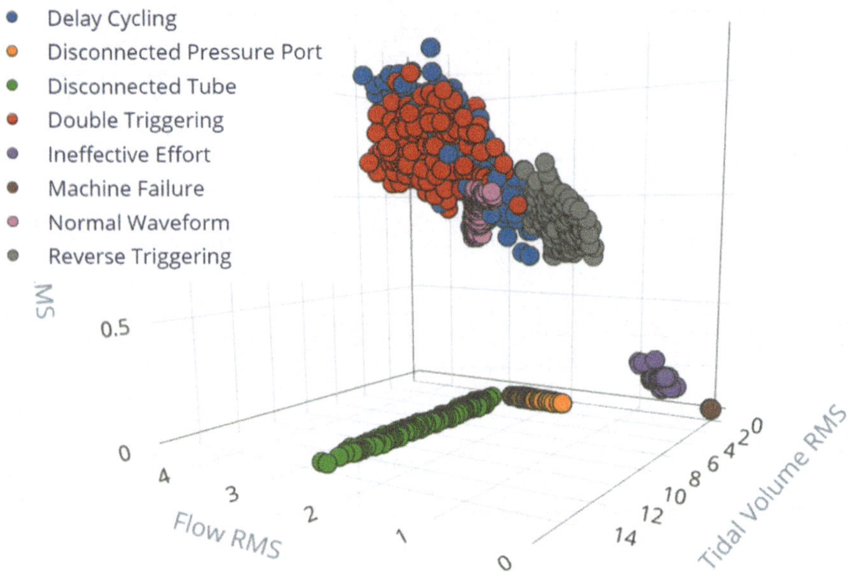

Fig. 4 3D graph of features extracted from the raw data of the emulated ventilator waveforms

extraction is a dimensionality reduction technique that reduces a large set of raw data into smaller groups for processing while retaining the information in the original data set [11, 12]. Analyzing time-series signals such as sensor data from the ventilator, this study employed spectral analysis as a features extraction algorithm. It processes the ventilator time-series signal to convert it into a frequency domain that extracts its spectrum characteristics. Figure 4 shows a 3D graph of RMS features extracted from the raw data of the emulated ventilator waveforms. The spectral analysis was able to group each class, making it easier for the ML algorithm to generalize the data. The algorithm extracted 11 spectral features of the raw data per axis; there were 33 features as input to the Neural Network classifier.

3.5 Ventilator Asynchrony Recognition Model and Near Cloud System

This study employed the optimized deep neural network enough to run on micro-controllers that classify the deviation of time-series waveform signal from the normal operation in real-time and standalone. Microcontrollers have limited memory and processing power, which places constraints on the sizes of machine learning models. The model was trained through the TensorFlow-based AutoML platform and converted the final model into the TensorFlow Lite version which allowed running the model on a microcontroller. The researchers chose ESP32-based processors (e.g., DOIT DevKit V1) to combine AI/ML capability with its IoT applications.

The researchers developed a data caching system, a wireless mesh network called AIC Near/Mobile Cloud, a private cloud infrastructure that was also included in some projects i.e., the A-vent, and a phototherapy light system for jaundice treatment that allows IoT devices to communicate [5, 13, 14]. The local data caching system can collect real-time data from sensor-equipped medical machines and perform real-time data analysis for hospital medical staff on multiple machines. The device's server connects the IoT medical machines to the time-series database that can store real-time data and analysis performed by AI/ML. It automatically stores the data in Unix timestamp format that allows graphing the historical clinical data with descriptive analytics in the remote monitoring dashboard for clinician reference.

4 Results and Discussions

4.1 Ventilator Asynchrony Recognition Alarms

The team employed the TinyML approach to classifying various types of asynchrony beyond the normal waveform that generates an alert. The results were obtained from 30 to 50 s of breath cycling, where the asynchronies and ventilator circuit status were emulated after 3 normal breath cycling. Figure 5 shows the alarms for emulated PVA. The breath cycling consists of pressure, flow, and volume represented by blue, orange, and green lines, respectively. The alarm signal is represented by a red line, if its amplitude is high, the sequence of positive alarms is detected to generate an alert signal.

It shows the delay cycling (DC), double triggering (DT), ineffective effort (IE), and reverse triggering (RT) asynchronies. The embedded neural network was able to recognize the asynchronies from normal waveforms in near real-time. The basis of alarm is the prediction accuracy and its labels. The positive alarm event is described if the predicted breath cycling is other than the normal waveform, and when the prediction accuracy surpasses the confidence level threshold of 0.80. The waveforms are sampled at 50 Hz with 5 sequence samples of positive alarm to avoid triggering of false alarm. The alarm algorithm took ~6–15 s (3–7 breath cycles) to trigger the alarm signal. It only took 1–2 normal breath cycles to reset the alert signal.

Figure 6 shows the machine, ventilator circuit, and anomaly alarms that include the disconnected pressure port (DPP), disconnected tube (DT), machine failure (MF), and emulated anomalous asynchrony. The alarm algorithm took ~10–25 s to generate an alert signal, which is somehow longer for asynchrony alarms. Furthermore, it took ~2–12 s (1–6 breath cycle) to reset the alert signal. The anomalous waveform was taken by rapidly turning around the air release valve from side to side. The anomaly detection took ~4 s (2 anomalous breath cycles) to trigger the alarm signal when the anomaly score exceeds the normal threshold. It took ~10 s (5 normal breath cycles) to

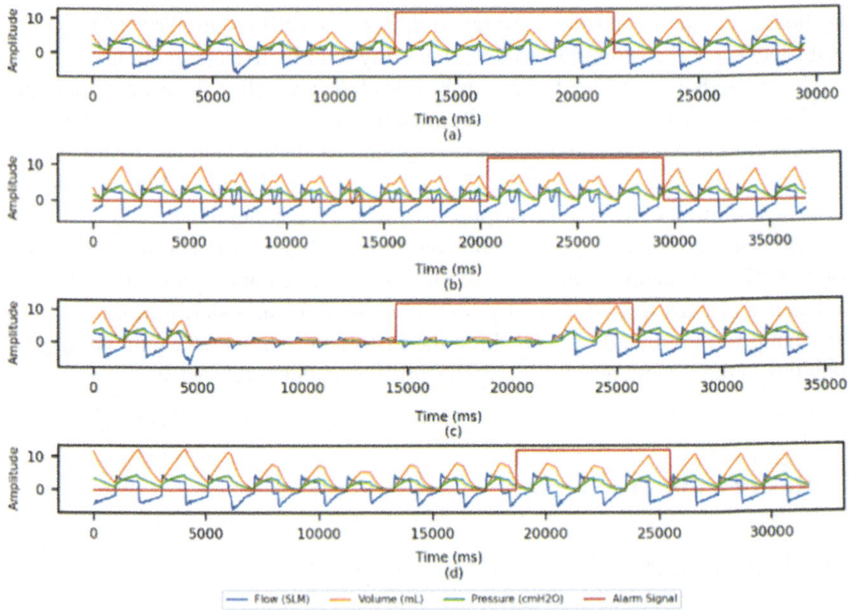

Fig. 5 Ventilator asynchrony alarms: **a** delay cycling, **b** double triggering, **c** ineffective effort, **d** reverse triggering

reset the alarm signal. Hence, this study proves that the future mechanical ventilator device can detect time-series types of alarms in near real-time, which assists the healthcare workers to reduce their workload and sustain the critical services of the healthcare system.

4.2 Ventilator Asynchrony Recognition Model Performance

To evaluate the model, the researchers randomly divided the dataset into (a) training, (b) validation, and (c) testing sets with 60%, 20%, and 20% partitions respectively. The model was evaluated using 10 k-fold cross-validations. The model performance for validation and testing sets was summarized using a confusion matrix as provided in Tables 2 and 3. It consists of m rows and n columns, where m is the actual emulated asynchrony and n is the asynchrony predicted by the algorithm. The diagonal elements show the accuracy of the predicted breath cycling matched with the actual emulated waveforms.

The weighted model accuracy resulted from validation and test sets are 97.8% and 98.01%, respectively. Both accuracies are relative to each other, thus the PVA and mechanical state recognition model can well generalize the emulated ventilator waveforms. The performance of the classifier model reflects how the features were

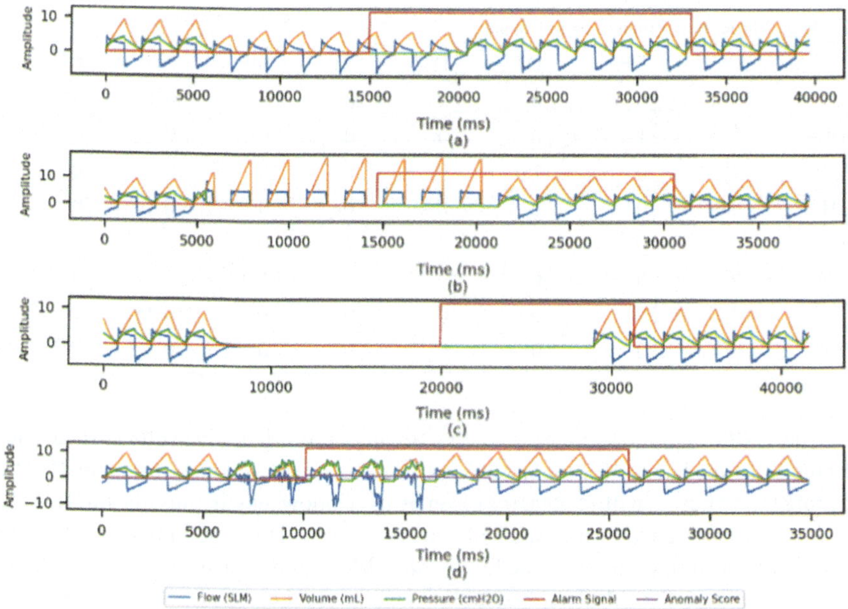

Fig. 6 Machine, patient airway circuit, and anomaly-related alarms: **a** disconnected pressure port, **b** disconnected tube, **c** machine failure, **d** anomalous asynchrony

Table 2 Model performance from validation set

	DC[1]	DPP[2]	DT[2]	DT[1]	IE[1]	MF[2]	N	RT[1]
DC[1]	0.93	0	0	0.04	0	0	0.01	0.02
DPP[2]	0	1.0	0	0	0	0	0	0
DT[2]	0	0.01	0.99	0	0	0	0	0
DT[1]	0.08	0	0	0.91	0	0	0.01	0
IE[1]	0	0	0	0	1.0	0	0	0
MF[2]	0	0	0	0	0	1.0	0	0
N	0	0	0	0.1	0	0	0.99	0
RT[1]	0	0	0	0	0	0	0	1.0

grouped as shown in Fig. 4. The delay cycling and double triggering overlapped each other causing confusion between them. The features for normal waveform and reverse triggering are concentrated. However, the data points were plotted near delay cycling and double triggering which has an insignificant effect on its performance. The rest of the classes were clustered independently which enabled the ML algorithm to be able to generalize data easily.

Table 3 Model performance from test set

	DC[1]	DPP[2]	DT[2]	DT[1]	IE[1]	MF[2]	N	RT[1]
DC[1]	0.90	0	0	0.02	0	0	0.01	0
DPP[2]	0	1.0	0	0	0	0	0	0
DT[2]	0	0	1.0	0	0	0	0	0
DT[1]	0.04	0	0	0.95	0	0	0	0
IE[1]	0	0	0	0	1.0	0	0	0
MF[2]	0	0	0	0	0	1.0	0	0
N	0	0	0	0	0	0	0.99	0
RT[1]	0	0	0	0	0	0	0	0.99

The ESP32 was able to process the stream of ventilator waveform data with 19 ms and 1 ms latency for the features extraction and inferencing, respectively. The features were buffed within 2000 ms given by its window length; thus, the inferencing results were printed after the data had been processed. This result proves that future ventilator machines can be embedded with AI/ML to detect time-series alarms in near real-time and stand-alone assist clinicians in monitoring critical patients.

5 Conclusion

As we embrace a circular economy for the development of biomedical devices, this study demonstrated a low-cost solution to upgrading medical machines such as ventilators with new AI/ML analysis and real-time data storage in the Near Cloud network. We modeled the patient-ventilator interaction by varying the airflow within the ventilator unit. The captured ventilator waveform was validated by the physician, as a proof of concept. The AIC team further improved the functionalities of the previous minimum viable ventilator by integrating a standalone alarm system utilizing embedded deep learning for near real-time detection of ventilator asynchrony and machine status, and clustering for detecting anomalies to assist clinicians in monitoring patients who require respiratory support. This development demonstrates how a conventional ventilator can be improved and linked to a new generation of medical machines/devices.

This study proved that the future mechanical ventilator machine can detect time-series types of alarms in near real-time, which assists the healthcare workers to reduce their workload and sustain the critical services of the healthcare system. Its AI predictive capabilities are supposed to support physicians in decision-making, not replace their expertise. The team also presented how the system can be integrated into the AIC Near/Mobile Cloud with the multiple sensor-equipped medical machines as part of the IoMT system initiatives.

Acknowledgements We thank Dr. Emma Porio and the Coastal Cities-at-Risk in the Philippines for the initial funding of the team's efforts from the A-vent up to the sensors and the University Research Council of the Ateneo de Manila University for granting financial support to continue the work. We thank the Lord for this opportunity to work on a life-saving device amidst the limitations brought about by the pandemic.

To the DOST-SEI ERDT program which provides a full scholarship and research grant for graduate students. To the Electronics, Computer, and Communication Engineering department, we express our very great appreciation for the valuable support, advice, and guidance for this study.

References

1. King WP, et al (2020) Emergency ventilator for COVID-19. PLoS One 15(12). https://doi.org/10.1371/JOURNAL.PONE.0244963
2. Grimm CA (2020) Hospital experiences responding to the COVID-19 pandemic: results of a national pulse survey march 24–27, 2020 (OEI-06-20-00300; 04/20). Accessed Mar. 14, 2022. [Online]. Available: oig.hhs.gov/oei/reports/oei-06-20-00300.asp
3. Gholami B, Haddad WM, Bailey JM (2018) AI in the ICU: in the intensive care unit, artificial intelligence can keep watch
4. Rackley CR (2020) Monitoring during mechanical ventilation. Respir Care 65(6):832–846. https://doi.org/10.4187/RESPCARE.07812
5. Cabacungan PM, et al (2021) Design and development of A-vent: a low-cost ventilator with cost-effective mobile cloud caching and embedded machine learning. In: TENSYMP 2021—2021 IEEE region 10 symposium. https://doi.org/10.1109/TENSYMP52854.2021.9550920
6. Williams LM, Sharma S (2021) Ventilator safety. StatPearls. Accessed Mar. 14, 2022. [Online]. Available: https://www.ncbi.nlm.nih.gov/books/NBK526044/
7. Corey RM et al (2020) Low-complexity system and algorithm for an emergency ventilator sensor and alarm. IEEE Trans Biomed Circuits Syst 14(5):1088–1096. https://doi.org/10.1109/TBCAS.2020.3020702
8. Mashoedah et al (2021) IoT enabled ventilator monitoring system for Covid-19 patients. J Phys: Conf Ser 2111(1):012035. https://doi.org/10.1088/1742-6596/2111/1/012035
9. Rehm GB et al (2020) Leveraging IoTs and machine learning for patient diagnosis and ventilation management in the intensive care unit. IEEE Pervasive Comput 19(3):68–78. https://doi.org/10.1109/MPRV.2020.2986767
10. Tripathy S, Sahoo L (2015) A survey of different methods of clustering for anomaly detection. Int J Sci Eng Res 6(1). Accessed Jan. 10, 2022. [Online]. Available: https://www.ijser.org/paper/A-Survey of different methods of clustering for anomaly detection.html
11. Feature Extraction - MATLAB & Simulink. Accessed Jan. 08, 2022 https://www.mathworks.com/discovery/feature-extraction.html
12. Feature Extraction Definition | DeepAI. Accessed Jan. 08, 2022 https://deepai.org/machine-learning-glossary-and-terms/feature-extraction
13. Talusan JP, Nakamura Y, Mizumoto T, Yasumoto K (2018) Near cloud: low-cost low-power cloud implementation for rural area connectivity and data processing. In: Proceedings-international computer software and applications conference, vol 2, pp 622–627, https://doi.org/10.1109/COMPSAC.2018.10307
14. Cabacungan PM, Oppus CM, de Guzman JE, Tangonan GL, Culaba IB, Cabacungan NG (2019) Intelligent sensors and monitoring system for low-cost phototherapy light for jaundice treatment. In: 2019 International symposium on multimedia and communication technology, ISMAC 2019. https://doi.org/10.1109/ISMAC.2019.8836133

Effect of Fluid Visco-Temperature Properties on Pilot-Operated Relief Valve Stability

Dong Wang, Yaobao Yin, Jiayang Yuan, Junyong Fu, and Wending Li

Abstract The viscosity of fluid varies with temperature, and the dynamic behavior of the relief valve may be impacted in this situation. After obtaining the oil visco-temperature properties, the dynamic mathematical model of a pilot-operated relief valve is derived. To obtain the effect of the visco-temperature properties on the stability of pilot-operated relief valves, the system transfer function block diagram is constructed. The frequency-domain analysis method was then utilised to investigate its dynamic behavior. The results suggest that the viscosity of the oil is thin at high temperatures, resulting in low flow resistance in the dynamic feedback orifice. In this case, the mass-spring vibration system composed of the main poppet and mechanical spring has a natural frequency close to the relief valve's operating frequency, which is not conducive to the stability of the relief valve. The oil inside the main valve spring chamber, however, is equivalent to a liquid spring with an extraordinarily high stiffness since the viscosity of the fluid is thicker at low temperatures, which decreases the flow capacity of the orifice. The main valve's mass spring vibration system has a natural frequency in this instance that is significantly higher than the valve's operating frequency, enhancing the stability of the relief valve. Finally, the numerical simulation results demonstrate the rationality of the theoretical analysis.

Keywords Pilot-operated relief valve · Stability · Frequency-domain analysis method · Visco-temperature properties

D. Wang · Y. Yin (✉) · J. Yuan
Tongji University, Shanghai City 201804, P.R. China
e-mail: y-yin@tongji.edu.cn

J. Fu · W. Li
Shanghai Aerospace Control Technology Institute, Shanghai City 201109, P.R. China

G. L. Conte and O. Sename (eds.), *Proceedings of the 11th International Conference on Mechatronics and Control Engineering*, Lecture Notes in Mechanical Engineering,
https://doi.org/10.1007/978-981-99-6523-6_11

1 Introduction

The relief valve is a basic pressure control component in hydraulic circuits and is widely used to hold a preset pressure range in the system or to protect circuits from overloading [1]. The pilot-operated relief valve is a two-stage structure that was invented by Vickers in 1931 [2]. Because of their excellent pressure-flow characteristics, pilot-operated relief valves are frequently employed in high-pressure and large-flow conditions. Inevitably, the relief valve always encounters the problem of switching between extremely high and extremely low temperatures. It is worth noting that the viscosity of oil changes with temperature, which in turn affects the dynamic characteristics of the hydraulic system. Scholars have carried out a lot of research on the viscosity-temperature characteristics of fluids. Reynolds equation, Slotte equation, Erying equation, Walther equation, and others have been proposed [3, 4]. In addition, the researchers studied the influence of fluid visco-temperature properties on the stability of hydraulic systems, and several improving methods were put forward. The most efficient of these is the method of thickening low-viscosity oils using high-polymer additives [5]. The vibration phenomenon of the relief valve has also attracted great interest from researchers. Hayashi demonstrated that a variety of factors, including the impact of the spool and seat, the hysteresis of transient hydrodynamic force, the piping system, and the coupling phenomenon between components, can induce poppet valve instability [6]. Besides, many studies show that the spool motion damping coefficient, orifice size, spool half cone angle, pre-compression of the adjusting spring, and valve chamber volume all affect relief valve stability [7–18]. Merritt analyzed the stability of a relief valve and concluded that a good match between fixed throttle and sensitive chamber volume can improve the stability of the relief valve [19].

However, there is no convincing explanation for the instability of the pilot relief valve caused by the visco-temperature properties of the oil. In this research, the effect of visco-temperature properties on pilot-operated relief valve stability is investigated. This article is organized as follows: After describing the operating principle of the studied valve, a detailed mathematical model is proposed. The effect of viscosity changes on the stability of a pilot-operated relief valve is further investigated by a frequency domain analysis method. Ultimately, numerical simulation was used to validate the theoretical conclusions.

2 Description of the Pilot-Operated Relief Valve

Typically, the pilot-operated relief valve consists of a main valve and a pilot valve. Figure 1 illustrates the structure of a pilot-operated relief valve, and the corresponding schematic diagram is shown in Fig. 2.

As shown in Fig. 2, The supply pressure from chamber A is transmitted to chamber C via the sharp-edged orifice R_1 and acts on the pilot valve, then to chamber B via

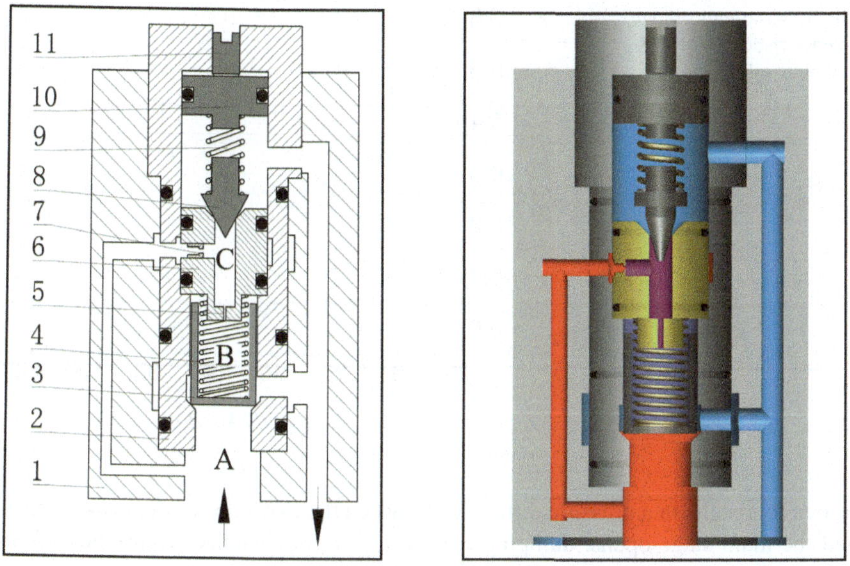

Fig. 1 Structure diagram of a pilot-operated relief valve: 1. Valve block; 2. Valve sleeve; 3. Main poppet; 4. Mainspring; 5. Dynamic pressure feedback orifice R_2; 6. Pilot valve seat (mainspring seat); 7. Sharp-edged orifice R_1; 8. Pilot poppet; 9. Pilot spring; 10. Pilot spring seat; 11. Adjustment bolt; A. Pressure controlled chamber; B. Pressure sensing chamber; C. Pressure sensitive chamber

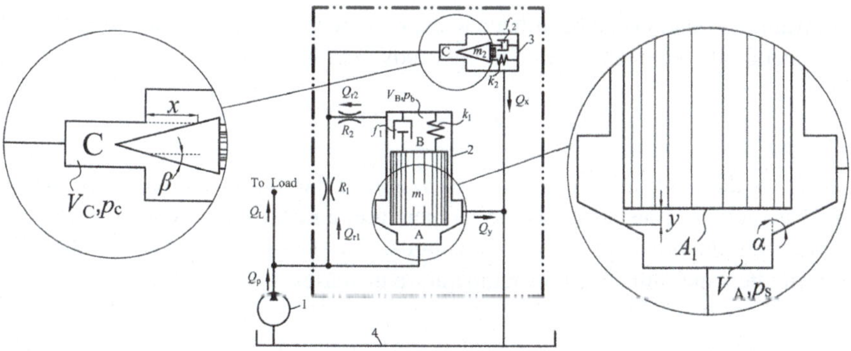

Fig. 2 Schematic diagram of a pilot-operated relief valve: 1. Hydraulic pump; 2. Main valve; 3. Pilot valve; 4. Tank

the dynamic pressure feedback orifice R_2. The pilot valve remains closed by preload spring force as long as the pressure p_c does not exceed the value preset at the pilot stage. Meanwhile, Pascal's Law holds that $p_s = p_b = p_c$. The main poppet is in a hydraulically balanced state, but it is also seated on the valve seat by preloaded spring force. Once the system pressure p_s reaches the so-called cracking pressure, the pilot valve opens to release pressure in chambers C and B. The pressure loss caused by the

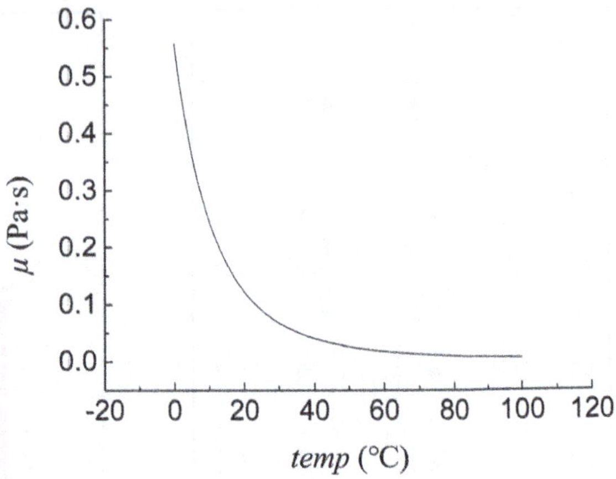

Fig. 3 Visco-temperature characteristics of the 46# hydraulic oil

orifice R_1 results in a pressure difference between the two sides of the main poppet, and the main stage opens, dumping excess fluid from chamber A into the tank to achieve system pressure regulation.

3 Visco-Temperature Properties of the Fluid

The Walther Equation can be utilized to express the visco-temperature properties of the mineral oil type medium used in this study (46# hydraulic oil).

$$\lg \lg(\nu + N) = L + M \lg(273 + t) \tag{1}$$

where ν is the kinematic viscosity, $\nu = \mu/\rho$, μ is the dynamic viscosity, L and M are characteristic constants, t is the temperature, and N is a general constant that applies to all fluids except low viscosity fluids, $N = 0.7$. As shown in Fig. 3, the oil's viscosity becomes thinner as the temperature increases.

4 Mathematical Model

The following assumptions were made in the formulation of the relief valve mathematical model: 1. The effective bulk modulus of the fluid is assumed to be constant. 2. The pressure at the relief valve outlet is assumed to be equal to the tank pressure. 3. Perturbations are assumed to be minimal so that the orifice characteristics can be linearized around the operating point. 4. Leakage around the main poppet is negligible.

4.1 Nonlinear Dynamic Mathematical Model

The dynamic mathematical model of the pilot-operated relief valve is as follows:

The flow passing through the relief valve is given by

$$Q_p - Q_L = Q_x + Q_y \tag{2}$$

where Q_p is the supply flow rate, Q_L is the flow rate to load, Q_x is the flow rate at the pilot port, and Q_y is the flow rate at the main port.

As shown in Fig. 1, when the flow continuity equation is applied to the chamber A of volume V_A, yields

$$Q_p - Q_L - Q_{r1} - Q_y - \frac{V_A}{E}\frac{dp_s}{dt} - A_1\frac{dy}{dt} = 0 \tag{3}$$

where Q_{r1} is the flow rate passing through the sharp-edged orifice R_1, E is the bulk stiffness of fluid, p_s is the pressure in chamber A, A_1 is the cross-sectional area of the main poppet, y is the main valve displacement.

The flow rate passing through the main exit port can be represented by the well-known relation

$$Q_y = C_{d1} A_y \sqrt{\frac{2p_s}{\rho}} \tag{4}$$

where C_{d1} is the discharge coefficient of the main exit port, A_y is the flow area of the main port, $A_y = \pi d_1 y\sin\alpha(1 - y\sin 2\alpha/(2d_1))$, d_1 is the diameter of the main exit port, α is the half-angle of the main valve.

The flow rate passing through the sharp-edged orifice R_1 is given by

$$Q_{r1} = C_{r1}\frac{\pi d_{r1}^2}{4}\sqrt{\frac{2(p_s - p_c)}{\rho}} \tag{5}$$

where C_{r1} is the discharge coefficient of orifice R_1, d_{r1} is the diameter of orifice R_1, and p_c is the pressure in chamber C.

The force equilibrium equation of the main poppet is expressed as

$$m_1\frac{d^2y}{dt^2} = A_1(p_s - p_b) - k_1(y + y_0) - C_{d1}C_{v1}\pi d_1 \sin(2\alpha)yp_s - f_1\frac{dy}{dt} \tag{6}$$

where m_1 is the effective main poppet mass (including 1/3 spring mass), p_b is the pressure in chamber B, k_1 is the mainspring stiffness, y_0 is the constant pre-compression of the mainspring, C_{v1} is main port velocity coefficient, and f_1 is the viscous damping coefficient of the main poppet, $f_1 = \mu A_{v1}/\delta_1$, A_{v1} is the equivalent wetting area of the main poppet, and δ_1 is the clearance between the main poppet and valve body.

The continuity equation as applied to chamber C, of volume V_C, yields

$$Q_{r1} + Q_{r2} - Q_x - A_2 \frac{dx}{dt} - \frac{V_C}{E}\frac{dp_c}{dt} = 0 \quad (7)$$

where Q_x is the flow rate through the pilot stage, A_2 is the cross-sectional area of the pilot poppet, and x is the pilot valve displacement.

The flow rate passing through the pilot exit port can be represented by

$$Q_x = C_{d2} A_x \sqrt{\frac{2p_c}{\rho}} \quad (8)$$

where C_{d2} is the discharge coefficient of the pilot exit port, A_x is the flow area of the main port. $A_x = \pi d_2 x\sin\beta(1 - x\sin2\beta/(2d_2))$, d_2 is the diameter of the pilot exit port, and β is the half-angle of the pilot valve.

The force equilibrium equation of the pilot spool is expressed as:

$$m_2 \frac{d^2x}{dt^2} = A_2 p_c - k_2(x + x_0) - C_{d2} C_{v2} \pi d_2 \sin(2\beta) x p_c - f_2 \frac{dx}{dt} \quad (9)$$

where m_2 is the effective pilot poppet mass, k_2 is pilot spring stiffness, x_0 is the pilot spring pre-compression, C_{v2} is pilot port velocity coefficient and f_2 is the viscous damping coefficient of the pilot spool, $f_2 = \mu A_{v2}/\delta_2$, A_{v2} is the equivalent wetting area of the pilot poppet, δ_2 is the clearance between the pilot poppet and valve body.

The flow rate passing through the dynamic pressure feedback orifice R_2 is given by:

$$Q_{r2} = \frac{\pi d_{r2}^4}{128 \mu l_{r2}} (p_b - p_c) \quad (10)$$

where d_{r2} is the diameter of orifice R_2, l_{r2} is the length of orifice R_2.

The continuity equation as applied to the chamber B, of volume V_B, yields

$$A_1 \frac{dy}{dt} - Q_{r2} - \frac{V_B}{E}\frac{dp_b}{dt} = 0 \quad (11)$$

4.2 Linearization Model and System Block Diagram

Considering small perturbations for Eqs. (3, 4, 5, 6, 7, 8, 9, 10, and 11) near the operating point under rated flow, and taking Laplace transforms, a linearized model of the dynamic behavior of the valve was developed, as the following equations.

$$Q_{\mathrm{p}} - Q_{\mathrm{L}} - Q_{\mathrm{y}} - Q_{\mathrm{r1}} - \frac{V_{\mathrm{A}}}{E} p_{\mathrm{s}} s - A_1 y s = 0 \tag{12}$$

$$Q_{\mathrm{y}} = K_{\mathrm{A}} p_{\mathrm{s}} + K_{\mathrm{B}} y \tag{13}$$

$$Q_{\mathrm{r1}} = K_{\mathrm{C}}(p_{\mathrm{s}} - p_{\mathrm{c}}) \tag{14}$$

$$(A_1 - K_{\mathrm{D}} y_{\mathrm{x}}) p_{\mathrm{s}} - A_1 p_{\mathrm{b}} = \left(m_1 s^2 + f_1 s + k_1 + K_{\mathrm{D}} p_{\mathrm{sx}}\right) y \tag{15}$$

$$Q_{\mathrm{r1}} + Q_{\mathrm{r2}} - Q_{\mathrm{x}} - A_2 x s - \frac{V_{\mathrm{C}}}{E} p_{\mathrm{c}} s = 0 \tag{16}$$

$$Q_{\mathrm{x}} = K_{\mathrm{E}} p_{\mathrm{c}} + K_{\mathrm{F}} x \tag{17}$$

$$(A_2 - K_{\mathrm{G}} x_{\mathrm{x}}) p_{\mathrm{c}} = \left(m_2 s^2 + f_2 s + k_2 + K_{\mathrm{G}} p_{\mathrm{cx}}\right) x \tag{18}$$

$$Q_{\mathrm{r2}} = G_{\mathrm{r}}(p_{\mathrm{b}} - p_{\mathrm{c}}) \tag{19}$$

$$A_1 y s - Q_{\mathrm{r2}} - \frac{V_{\mathrm{B}}}{E} p_{\mathrm{b}} s = 0 \tag{20}$$

The physical meanings and expressions of K_{n} are shown in Table 1.

The system block diagram of the pilot-operated relief valve may be developed based on the above linearization model, by taking the flow rate at the inlet as input and the pressure of chamber A as output, as shown in Fig. 4. The physical meanings and calculation formulas of ω_{n} in Fig. 4 are shown in Table 2. From Fig. 4, viscosity affects the dynamic behavior of circuits 1, 2, 3, and 4, which are marked in the figure. Circuits 1 and 4 among them contain the same components, indicating the transfer function of the main valve mass-spring vibration system. Circuit 2 depicts the pilot valve pressure control chamber's transfer function. The transfer function

Table 1 Physical meanings and expressions of K_{n}

Sign	Physical meaning	Expressions
K_{A}	Flow pressure coefficient of main exit port	$K_{\mathrm{A}} = C_{\mathrm{d1}}\pi d_1 \sin\alpha y_{\mathrm{x}}\sqrt{1/2\rho p_{\mathrm{sx}}}$
K_{B}	Flow gain of main exit port	$K_{\mathrm{B}} = C_{\mathrm{d1}}\pi d_1 \sin\alpha\sqrt{2p_{\mathrm{sx}}/\rho}$
K_{C}	Hydraulic conductivity of orifice $\mathrm{R_1}$	$K_{\mathrm{C}} = C_{\mathrm{r1}}\pi d_{\mathrm{r1}}^2\sqrt{1/2\rho(p_{\mathrm{sx}} - p_{\mathrm{cx}})}/4$
K_{D}	Equivalent stiffness of steady hydrodynamic force of the main valve	$K_{\mathrm{D}} = C_{\mathrm{d1}} C_{\mathrm{v1}}\pi d_1 \sin(2\alpha)$
K_{E}	Flow pressure coefficient of pilot port	$K_{\mathrm{E}} = C_{\mathrm{d2}}\pi d_2 x_{\mathrm{x}} \sin\beta\sqrt{1/2\rho p_{\mathrm{cx}}}$
K_{F}	Flow gain of pilot port	$K_{\mathrm{F}} = C_{\mathrm{d2}}\pi d_2 \sin\beta\sqrt{2p_{\mathrm{cx}}/\rho}$
G_{r}	Fluid conduit of orifice $\mathrm{R_2}$	$G_{\mathrm{r}} = \pi d_{\mathrm{r2}}^4/(128\mu l_{\mathrm{r2}})$

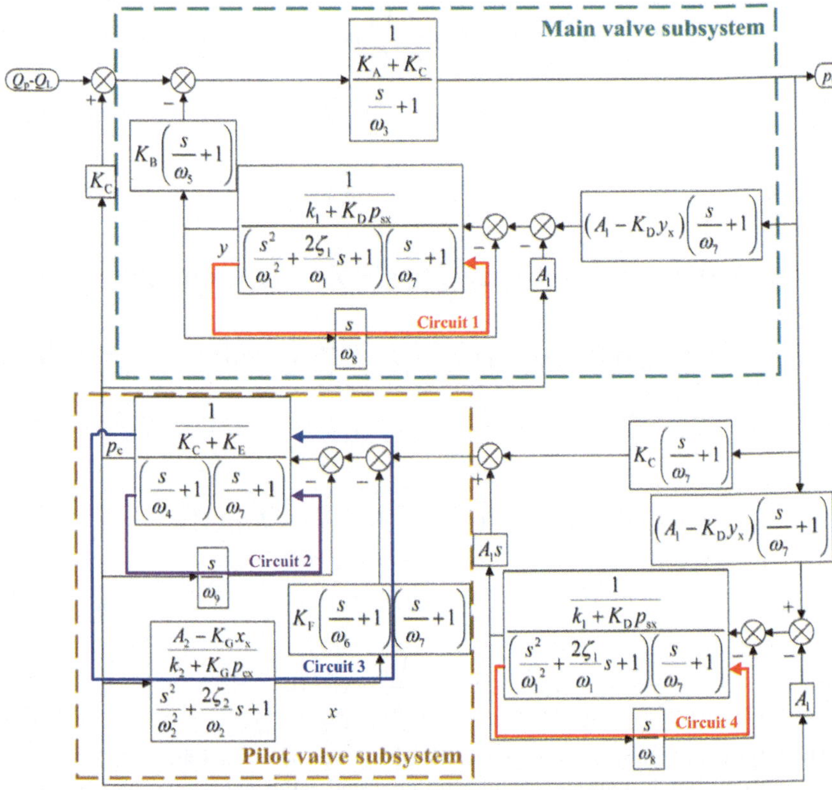

Fig. 4 The system block diagram of the pilot-operated relief valve

of the pilot valve spring-mass vibration system is represented by Circuit 3. The dynamic characteristics of both the main valve system and the pilot valve system will be impacted by the change in fluid viscosity, which can be demonstrated by the analysis above.

5 Frequency Domain Analysis of Valve Subsystems Stability

In this section, the frequency domain analysis method was used to study the valve's dynamic behavior. The following Table 3 lists the complete parameters utilized in the investigation. The steady-state operating point of the relief valve, under a flow rate of 220 L/min, 40 °C, can be determined by solving Eqs. (3, 4, 5, 6, 7, 8, 9, 10, and 11) through a simple analytical method, as listed in following Table 4.

Table 2 Physical meanings and expressions of ω_n

Sign ω_n	Physical meaning	Expressions
ω_1	Natural frequency of main spring-mass vibration system	$\omega_1 = \sqrt{(k_1 + K_D p_{sx})/m_1}$
ω_2	Natural frequency of pilot spring-mass vibration system	$\omega_2 = \sqrt{(k_2 + K_G p_{cx})/m_2}$
ω_3	Break-frequency of the chamber A	$\omega_3 = E(K_A + K_C)/V_A$
ω_4	Break-frequency of the chamber C	$\omega_4 = E(K_C + K_E)/V_C$
ω_5	Break-frequency of the differential link of main port	$\omega_5 = K_B/A_1$
ω_6	Break-frequency of the differential link of pilot port	$\omega_6 = K_F/A_2$
ω_7	Break-frequency of the chamber B	$\omega_7 = EG_r/V_B$
ω_8	Break-frequency of the orifice R_2	$\omega_8 = G_r/A_1^2$
ω_9	Break-frequency of the chamber B	$\omega_9 = V_B/E$

Table 3 Main parameters utilized in the investigation

Parameter	Value	Parameter	Value	Parameter	Value
m_1 (g)	30	α (°)	30	V_A (mL)	1000
m_2 (g)	5	β (°)	15	V_B (mL)	0.5
k_1 (N/mm)	20	A_{v1} (mm^2)	220	V_C (mL)	0.7
k_2 (N/mm)	50	A_{v2} (mm^2)	94.2	l_1 (mm)	25
x_0 (mm)	3.5	δ_1 (μm)	10	l_2 (mm)	10
y_0 (mm)	10.8	δ_2 (μm)	20	d_{r1} (mm)	1.0
d_1 (mm)	20	C_{d1}, C_{d2}	0.65	d_{r2} (mm)	1.4
d_2 (mm)	3	C_{r1}	0.61	l_{r2} (mm)	10
E (MPa)	800	ρ (kg/m^3)	889	μ (Pa·s)	0.041 (40 °C)

Table 4 Steady-state operating point

Parameter	Value	Parameter	Value
Chamber A pressure p_{sx}(MPa)	30.0	Pilot valve displacement x_x (mm)	0.11
Chamber C pressure p_{cx}(MPa)	27.1	Main valve displacement y_x (mm)	0.68

5.1 Stability of Pilot Valve Subsystem

The system transfer function of the pilot valve subsystem is

$$G_{\text{pilot}}(s) = \frac{\frac{K_F(A_2 - K_G x_x)}{(K_C + K_E)(k_2 + K_G p_{cx})}\left(\frac{s}{\omega_7} + 1\right)\left(\frac{s}{\omega_6} + 1\right)}{\left\{\frac{s^2}{\omega_4\omega_7} + \left[\frac{1}{\omega_4} + \frac{1}{\omega_7} + \frac{1}{(K_C + K_E)\omega_9}\right]s + 1\right\}\left(\frac{s^2}{\omega_2} + \frac{2\zeta_2}{\omega_2}s + 1\right)} \quad (21)$$

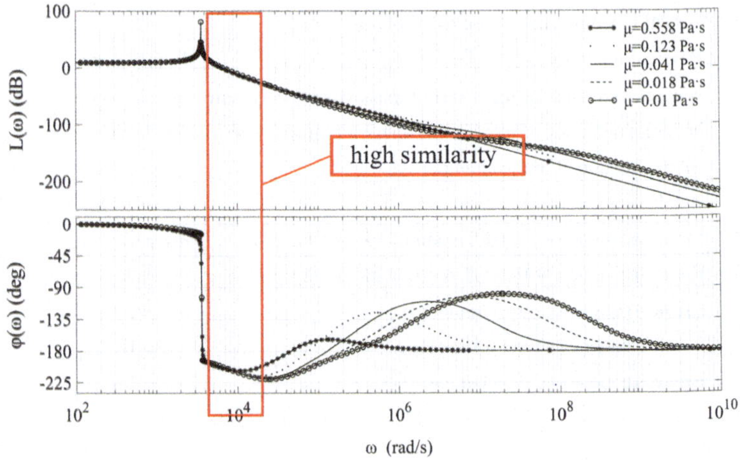

Fig. 5 The Bode diagram of pilot valve subsystem $G_{\text{pilot}}(s)$ at various viscosities

Figure 5 depicts the Bode diagrams of the pilot valve subsystem at various viscosities. There is not much variation in the dynamic characteristics because the pilot valve subsystem's dynamic characteristics are virtually unaffected by changes in viscosity.

5.2 *Stability of Main Valve Subsystem*

The system transfer function of the main valve subsystem is

$$G_{\text{main}}(s) = \frac{\frac{K_B(A_1 - K_D y_x)}{(K_A + K_C)K_{m1}}\left(\frac{s}{\omega_5} + 1\right)\left(\frac{s}{\omega_7} + 1\right)}{\left[\frac{s^3}{\omega_1^2\omega_7} + \left(\frac{2\zeta_1}{\omega_1\omega_7} + \frac{1}{\omega_1}\right)s^2 + \left(\frac{1}{\omega_7} + \frac{2\zeta_1}{\omega_1} + \frac{1}{K_{m1}\omega_8}\right)s + 1\right]\left(\frac{s}{\omega_3} + 1\right)} \tag{22}$$

where K_{m1} is the main valve equivalent mechanical spring stiffness, $K_{m1} = k_1 + K_D p_{sx}$.

Figure 6 depicts the Bode diagrams of the main valve subsystem at various viscosities.

The main valve displays various dynamic behaviors depending on the viscosity grade. The higher the viscosity, the better the stability margin of the main valve. But it's also clear that there are two situations. In the case of low viscosity, an oscillation element and a first-order inertial element constitute the characteristic equation of the main valve system. The low viscosity resulting in low flow resistance in the orifice R_2. In this case, the mass-spring vibration system composed of the main poppet and mechanical spring has a natural frequency close to the operating frequency of the relief valve, which is not conducive to the stability of the relief valve. However, the through-flow capacity of the orifice R_2 is severely decreased if the viscosity reaches a particular level ($\mu = 0.0558$ Pa·s). The oil inside chamber B is equivalent to a

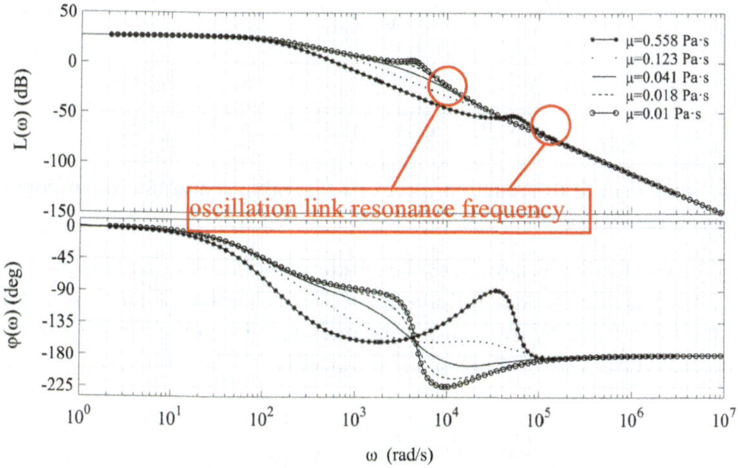

Fig. 6 The Bode diagram of main valve subsystem $G_{\text{main}}(s)$ at various viscosities

liquid spring with an extraordinarily high stiffness. The resonance frequency of the oscillation element in the main valve control system is significantly higher than the natural frequency of the main spring mass vibration system and considerably farther away from the operating frequency of the relief valve, and may improves the stability of the main valve subsystem.

Based on the above analysis, it can be concluded that: if the temperature of the fluid is low enough, a larger viscosity value may effectively enhance the stability of the main valve subsystem. And local stability like this may have a positive effect on the global stability of the pilot-operated relief valve.

6 Nonlinear Simulation of the Valve Under Study

The pilot-operated relief valve has 3 operational states.

I. As oil injection, p_s and p_c gradually rise, but the system pressure p_s is too low to open the main valve and pilot valve at this time. There is no flow at both the main and pilot ports, $Q_y = 0$, $Q_x = 0$. The mathematical model of this state is

$$Q_p - Q_L - Q_{r1} - \frac{V_A}{E}\frac{dp_s}{dt} = 0 \tag{23}$$

$$A_1(p_s - p_b) - k_1(y + y_0) + F_{t1} = 0 \tag{24}$$

$$Q_{r1} + Q_{r2} - \frac{V_C}{E}\frac{dp_c}{dt} = 0 \tag{25}$$

$$Q_{r2} + \frac{V_B}{E}\frac{dp_b}{dt} = 0 \tag{26}$$

$$A_2 p_c - k_2(x + x_0) + F_{t2} = 0 \tag{27}$$

where F_{t1} and F_{t2} are the reaction forces of the valve seat to the main poppet and pilot poppet, respectively.

II. p_s and p_c rise further as a result of the continuous oil inflow, and the pilot valve is opened. However, the pressure differential between chambers A and B is insufficient to open the main valve by overcoming the main valve's spring force. Currently, the pilot-operated relief valve's mathematical model is

$$Q_p - Q_L - Q_{r1} - \frac{V_A}{E}\frac{dp_s}{dt} = 0 \tag{28}$$

$$0 = A_1(p_s - p_b) - k_1(y + y_0) + F_{t1} \tag{29}$$

$$Q_{r1} + Q_{r2} - Q_x - A_2\frac{dx}{dt} - \frac{V_C}{E}\frac{dp_c}{dt} = 0 \tag{30}$$

$$Q_{r2} + \frac{V_B}{E}\frac{dp_b}{dt} = 0 \tag{31}$$

$$m_2\frac{d^2x}{dt^2} = A_2 p_c - k_2(x + x_0) - C_{d2}C_{v2}\pi d_2 \sin(2\beta) x p_c - f_2\frac{dx}{dt} \tag{32}$$

III. Once the p_s increases to the pre-set value, both the pilot valve and the main valve are open. In this case, the mathematical model of the pilot-operated relief valve is consistent with Sect. 3.1 (Eqs. 3, 4, 5, 6, 7, 8, 9, 10, and 11).

In this section, the dynamic numerical simulation model for such a valve was produced in the software Matlab/Simulink using solver ode45, which implements a Runge–Kutta method with a fixed time step(1×10^{-6}) for accurate computation. The supply flow (Q_p−Q_L), which steps from 0 to 220 L/min at 0.1 s, is taken as the input signal, and with the chamber A pressure p_s as an output signal, the dynamic characteristics of the relief valve can be derived. Figure 7 depicts the relief valve's three operational modes.

The system pressure p_s produced from numerical simulation with varied viscosities are shown in Fig. 8. The spool motion damping may be relatively modest if the fluid viscosity is low ($\mu = 0.0096$ Pa·s), which might lead to instability of the relief valve. However, if the viscosity increases to a certain value ($\mu \geq 0.041$ Pa·s), the relief valve's stability showed a good state. Considering that the highly viscous fluid will significantly dampen the movement of the valve core. Moreover, it is worth noting that the system pressure will vibrate strongly for a while when the viscosity $\mu = 0.0018$ Pa·s, and then the amplitude will gradually diminish (Fig. 8b). This is

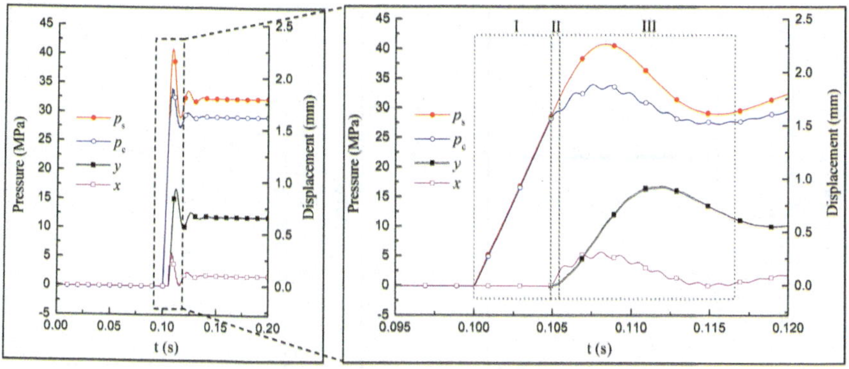

Fig. 7 Dynamic parameters of the pilot-operated relief valve obtained by simulation

because the pilot valve will impact the valve seat during the opening adjustment process as a result of the under-damping situation. As the system pressure vibration amplitude decreases, so does the pilot spool and main poppet vibration amplitude. The system pressure p_s won't fluctuation greatly once the pilot spool no longer hits the valve seat. Based on the above analysis, the simulation results are in agreement with the previous theoretical analysis. This means that the conclusion derived in Sect. 4 is reasonable (Fig. 9).

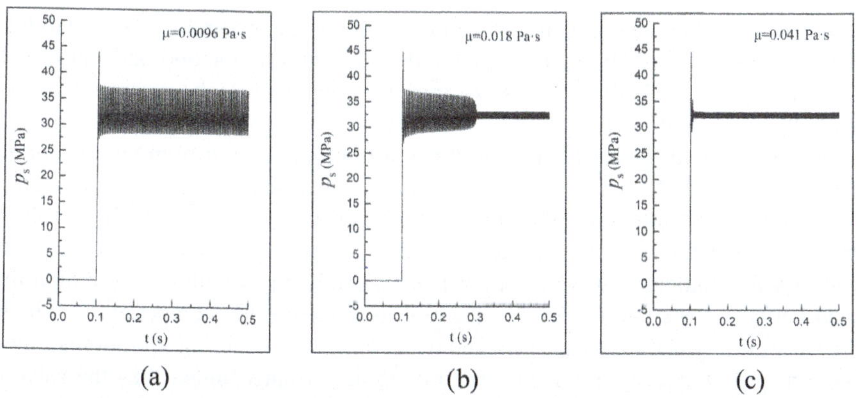

Fig. 8 System pressure p_s produced from numerical simulation with varied viscosities. **a** $\mu = 0.0096$ Pa·s **b** $\mu = 0.0018$ Pa·s **c** $\mu = 0.041$ Pa·s

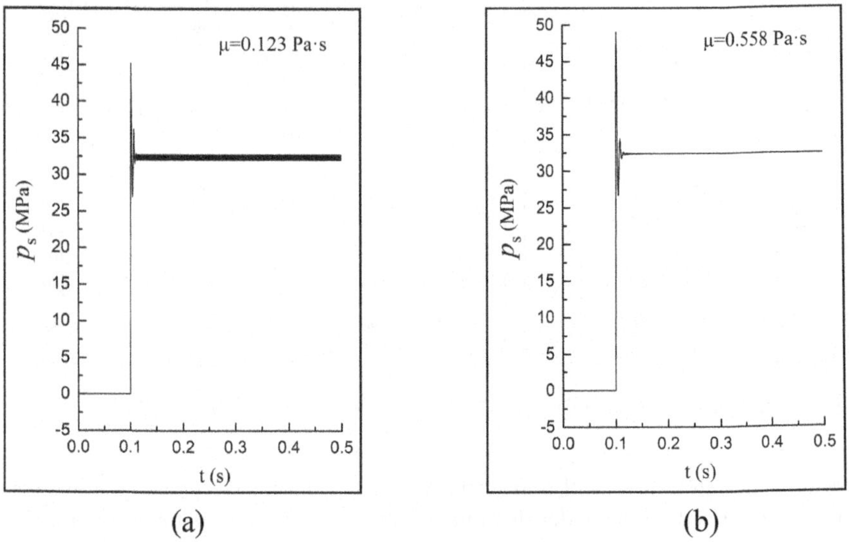

Fig. 9 System pressure p_s produced from numerical simulation with varied viscosities. **a** $\mu =$ 0.0123 Pa·s **b** $\mu = 0.558$ Pa·s

7 Conclusions

The effect of visco-temperature properties on pilot-operated relief valve stability has been investigated through frequency domain analysis method and numerical simulation. The analysis results show that, the pilot valve subsystem stability is virtually unaffected by changes in viscosity. However, the stability of the main valve system varies with viscosity. The fluid has a thin viscosity at high temperatures, resulting in a low flow resistance in the dynamic feedback orifice. In this case, the mass-spring vibration system has a natural frequency close to the valve's operating frequency, which is not conducive to the relief valve stability. The oil inside the main valve spring chamber, however, is equivalent to a liquid spring with an extraordinarily high stiffness since the viscosity of the fluid is thicker at low temperatures, which decreases the orifice flow capacity. The main valve's mass spring vibration system has a natural frequency in this instance that is significantly higher than the valve's operating frequency, enhancing the stability of the relief valve. Numerical simulations were produced to validate the above theoretical analysis, and excellent agreement between the simulation and the theoretical analysis was obtained. Simulation results show that the stability of the pilot-operated relief valve is poor at thin viscosity and better at high viscosity.

Acknowledgements The authors would like to acknowledge the support of National Natural Science Foundation of China (No. 52175059) and National Key Research and Development Program of China (2018YFB2001100).

References

1. Yaobao YIN (2019) Electro hydraulic control theory and its applications under extreme environment. Elsevier Inc, Oxford, pp 273–281
2. Harry VF (1931) Liquid relief valve. Patent US2043453, USA
3. Hatton RE (1962) Introduction to hydraulic fluids. Reinhold, New York
4. Jiang J, Ji T, Li Y et al (1998) Research on viscosity-temperature relationship of hydraulic oils. Lubr Eng 5:35–37
5. Firsov AP, Zaskalko PP, Kichkin GI, et al (1972) Mechanical stability of polymer-thickened oils used as working fluids for hydraulic systems of variable-pitch marine propellers. Chem Technol Fuels Oils:64–68
6. Hayashi S (1995) Instability of poppet valve circuit. JSME Int J 38(3):357–366
7. Dasgupta K, Karmakar R (2001) Dynamic analysis of pilot operated pressure relief valve. Simul Model Pract Theory 10:35–49
8. Maiti R, Saha R, Watton J (2002) The static and dynamic characteristics of a pressure relief valve with a proportional solenoid-controlled pilot stage. J Syst Control Eng 216:143–156
9. Johnston N, Edge KA, Brunelli M (2002) Impedance and stability characteristics of a relief valve. Proc Inst Mech Eng Part I J Syst Control Eng 216(5):371–382
10. Suzukia K, Urata E (2008) Dynamic characteristics of a direct pressure sensing water hydraulic relief valve. Int J Fluid Power 9(2):5–13
11. Qifang Y, Jiangping C (2009) Dynamic analysis of a pilot-operated two-stage solenoid valve used in pneumatic system. Simul Model Pract Theory 17:794–816
12. Moussou P (2009) Instability of pressure relief valves in water pipes. In: Proceedings of the ASME 2009 pressure vessels and piping division. Prague, Czech Republic, 26–30 July, pp 1–8
13. Yi D, Lu L, Zou J et al (2014) Interactions between poppet vibration and cavitation in relief valve. Proc Inst Mech Eng C J Mech Eng Sci 203:1989–1996
14. Min W, Wang H, Zheng Z et al (2020) Visual experimental investigation on the stability of pressure regulating poppet valve. Proc Inst Mech Eng C J Mech Eng Sci 234(4):2329–2348
15. Osama G (2020) Modeling of impact of the poppet element on its seat body in pressure relief valves. Proc Inst Mech Eng Part I J Syst Control Eng 235(5):703–720
16. Rao BS, Chandu DG (2014) Petrol engine exhaust valve design, analysis and manufacturing processes. Int J Mech Eng Rob Res 3(4):395–401
17. Pardeep R, Vineet K, Feras AH (2018) Fuzzy gain scheduling of an I-P-D controller for oscillation compensation in a sticky pneumatic control valve. Int J Mech Eng Rob Res 7(3):240–249
18. Jongha L, Kotaro T, Toshiharu K (2020) The impact of geometric design on energy efficiency of pneumatic booster valves with energy recovery. Int J Mech Eng Rob Res 9(5):665–672
19. Merrit H (1967) Hydraulic control systems. John Willy & Sons, New York, pp 363–374

Using Wheel Slippage for Improved Maneuverability of 4 Wheel Steering Vehicles

Erdem Ata and A. Bugra Koku

Abstract In the domain of land vehicles, the researches which focus on the subject of slippage mostly develop methods for detecting and avoiding slippage or compensating for slippage. Even when slippage is considered as a potential advantage, this is mostly researched for high speeds. This paper proposes a method that uses slippage as an advantage even at low speeds. This is achieved by using the nonholonomic properties of a 4WS (4 wheel steering) vehicle with independent front and rear drives. It is experimentally shown that controlled low speed slippage can be advantegous for certain tasks by comparing the vehicle behavior to a similar 2WS (2 wheel steering) vehicle which cannot leverage slippage.

Keywords 4WS · 2WS · AWS · Ackerman steering · Wheel slip · Drift · Maneuverability

1 Introduction

Autonomous ground vehicles (AGVs) can be used in crowded spaces such as offices and homes, or on challenging terrains. They can be used for carrying items in an office or in a warehouse and carrying out rescue missions on challenging terrains. Uneven terrains or unpaved off-road scenarios may require high maneuverability to carry out certain tasks.

There are different vehicle structures to achieve different amounts of maneuverability. The most common vehicle structure used in daily life is 2WS (2 wheel steering) cars. While adding more complexity to the vehicle structure, 4WS (4 wheel steering) cars provide higher maneuverability by allowing a smaller turning radius [8]. Both of these structures are designed to mostly avoid slippage and some researches assume no-slip conditions while modeling these vehicles [7]. Some outdoor vehicles such as

E. Ata (✉) · A. B. Koku
Department of Mechanical Engineering, Middle East Technical University, Ankara, Türkiye
e-mail: ataerdem@gmail.com

METU-ROMER, Middle East Technical University, Ankara, Türkiye

G. L. Conte and O. Sename (eds.), *Proceedings of the 11th International Conference on Mechatronics and Control Engineering*, Lecture Notes in Mechanical Engineering,
https://doi.org/10.1007/978-981-99-6523-6_12

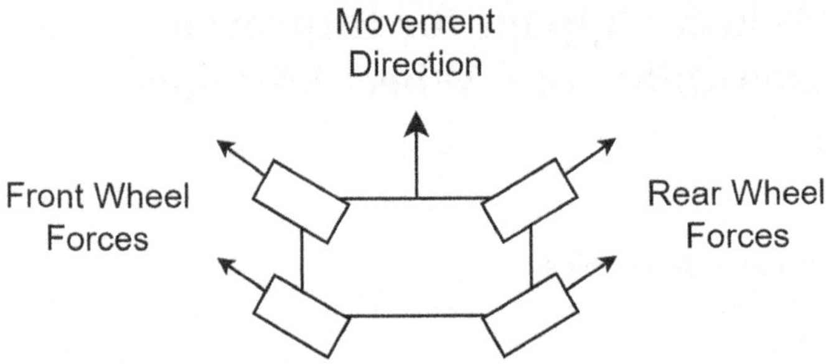

Fig. 1 An example case where the vehicle can move sideways using wheel slippage

tanks may utilize differential drives with no steering to provide arbitrarily small turning radii. However, all the movements that can be executed by differential drive vehicles involve slippage. The maneuverabilities of these vehicle structures have been extensively studied so far [5, 6, 10, 13, 14].

Wheel slippage is not desired for various reasons. On one hand slippage accumulates error in dead reckoning, on the other hand, it decreases controllability of vehicles. Hence, slippage in general is not a desired state. However, it may be useful to force these vehicles to slip for certain tasks. While vehicles with 4WS-4WD with independent front and rear drives and differential drives can start slippage at low speeds, a 2WS vehicle with Ackerman steering cannot.

In this paper, it is hypothesized and later shown that using 4WS with independent drives for front and rear wheels, one can control the vehicle to slip in an arbitrary direction without requiring prior maneuvers. As an example, consider the case shown in Fig. 1 where a 4WS vehicle steers at the opposite directions with the front and rear wheels while driving the front and rear wheels in the opposite directions. This actuation may cause the vehicle to drift sideways if controlled properly.

Although the results of excessive slippage may have adverse effects on wheels in everyday usage, using slippage only when it is completely necessary may allow cars to execute maneuvers that are otherwise harder or impossible such as fitting in tight parking spots or making very sharp maneuvers in crowded places. Since most 4WD electric cars already use independent electric motors for front and rear drives, the proposed structure is somehow similar to existing cars. However, controlling 4 inputs, 2 steering angles and 2 drive speeds, manually is a considerably more difficult task compared to driving a regular car. Therefore, the control of slippage needs to be automated for such applications in order to get the maximum benefit from wheel slippage.

In order to automate the controlled slippage to enhance maneuverability, we present a proper mathematical model for the proposed vehicle structure in the next section. While the starting point is the commonly used bicycle model, many additions and modifications are added to the model in order to make it more suitable for the case where the vehicle slips. The nonlinear model of the vehicle is then linearized to utilize linear state space methods.

Fig. 2 Small scale prototype

In the third section, two different control methods are discussed. The first one is a relatively naive approach that allows the user to directly use the coordinates of the vehicle as reference states using LQR. The second control method allows the user to control the vehicle to follow a certain path to reach to a given coordinate. In order to enable the vehicle to follow a certain path, a linear transformation is applied to the states of the system using the output matrix. Then, the controller's parameters are tuned in order to prioritize staying on the given line.

Finally, the proposed method is applied to the real life small scale prototype shown in Fig. 2. The feedback for the controller is provided using a Vicon optical motion tracking system. Using the same controller structure, a 2WS and a 4WS vehicle with independent drives (proposed structure) is compared. In order to compare the performances of different vehicles, their performance on following a relatively challenging trajectory is considered.

2 Mathematical Model

The mathematical model can be inspected in three parts: wheel forces, friction moment and input limits. After considering those points, a non-linear dynamic model is created using Kane's method and sympy mechanics module on Python. To utilize linear state space methods, the model is linearized around the instantaneous state at any time step using the Jacobian method. This requires the evaluation of system matrix and input matrix at each time step while keeping the linear approximation fairly accurate at any time.

Plant inputs are selected as steering angles (θ_f, θ_r) and angular velocities of the wheels (ω_f, ω_r). The states are selected as the position of the vehicle (x, y),

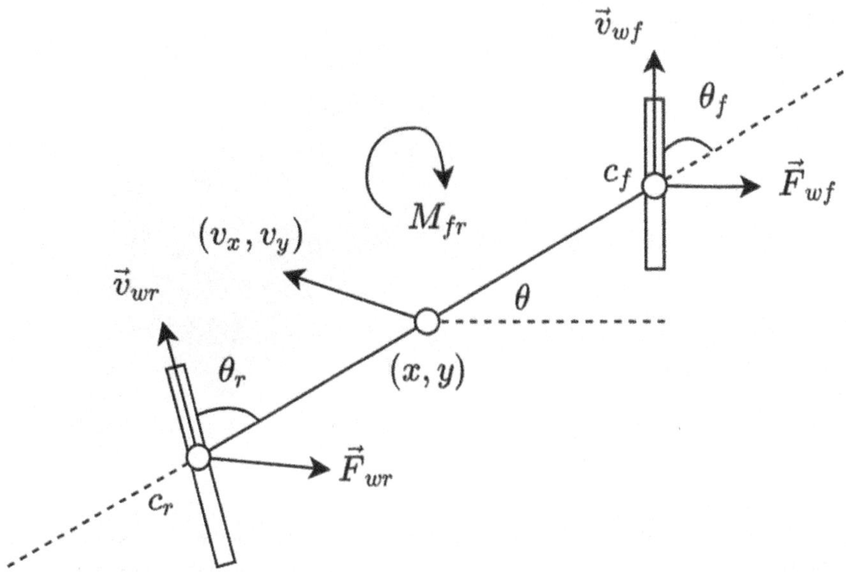

Fig. 3 Mathematical model

orientation of the vehicle with respect to the x axis (θ), the velocity of the vehicle (v_x, v_y) and the angular velocity of the vehicle (ω) respectively as shown in (1) and Fig. 3.

$$\mathbf{x} = \begin{bmatrix} x \\ y \\ \theta \\ v_x \\ v_y \\ \omega \end{bmatrix} \tag{1}$$

2.1 Wheel Forces

For the mathematical model, the vehicle is considered to be a rigid body in a planar space with 2 wheels (front and rear) attached as shown in Fig. 3. The friction between the wheels and the ground is assumed to be viscous. Although this assumption is clearly counter-intuitive, it results in a simple (i.e. less non-linear) mathematical model. It is hypothesized and observed in the experiments that the inaccuracy is not significant at tested speeds.

Under the viscous friction assumption, the force that is applied on the vehicle by a single wheel can be expressed with the following equation.

$$\vec{F}_{w,i} = c_i(\vec{v}_{w,i} - \vec{v}_{c,i}), \quad i = f, r \tag{2}$$

where c_i is the viscous friction coefficient between the wheel and the ground, $\vec{v}_w$ is the velocity which is dictated by the wheel and $\vec{v}_c$ is the velocity of the wheel center with respect to the ground.

For a given wheel, $\vec{v}_w$ can be expressed in terms of the inputs θ_i and ω_i.

$$\vec{v}_{w,i} = r\omega_i(\cos(\theta_i + \theta)\hat{\imath} + \sin(\theta_i + \theta)\hat{\jmath}), \quad i = f, r \tag{3}$$

where θ_i is the steering angle and ω_i is the angular velocity by which the wheel is driven. The subscript i is used to denote whether the wheel is front or rear.

2.2 Friction Moment

When the rotation of the vehicle around the front wheel alone is considered, the assumption of no friction due to the rotation can be acceptable. However, when there are 2 front wheels, in order to account for the friction caused by the rotation of the vehicle around the center of the front wheels, one has to introduce an extra moment to the bicycle model. This moment is assumed to be proportional to the angular velocity of the vehicle and the friction coefficient. It is also affected by the distance between the front wheels. Then the additional friction moment can be represented by the following expression.

$$\vec{M}_{fr} = -d\vec{\omega}(c_r + c_f) \tag{4}$$

where $\vec{M}_{fr}$ is the additional friction moment, d is the distance between the front wheels and $\vec{\omega}$ is the angular velocity of the vehicle.

2.3 Input Limits

Both the steering angle and the angular velocity of the wheels are limited to certain ranges. Due to the non-linearity of the system, simply clipping the controller outputs before feeding them into the plant would not work. The limits of these variables should be represented in the mathematical model.

In order to achieve that, 2 variables which are expressed with the following equations are introduced.

$$\omega_i = l_\omega \tanh(\omega_i^{initial}), \quad i = f, r \tag{5}$$

$$\theta_i = l_\theta \tanh(\theta_i^{initial}), \quad i = f, r \tag{6}$$

If the plant inputs are selected as $\omega_i^{initial}$ and $\theta_i^{initial}$, then the absolute values of steering angle and the angular velocity of the wheels will be limited to l_ω and l_θ respectively.

3 Control Method

Using the linear quadratic regulator (LQR) method, the states can be directly controlled to achieve a desired state. For a linear system, LQR method optimizes the expression given in (7). However, that does not put any constraint on the path that will be followed. For most tasks that require the vehicle to move to a certain location, a specific path should be followed. Using a proper output matrix, the vehicle can be controlled to follow a line with a specific speed. For more complicated paths, it is possible to approximate the path with line segments.

$$\int_0^\infty \mathbf{x}^T \mathbf{Q} \mathbf{x} + \mathbf{u}^T \mathbf{R} \mathbf{u} \, dt \tag{7}$$

For both methods, two important tricks are used. Firstly, since the controller is allowed to instantly change the steering angles in the mathematical model, this causes a discrepancy between the model and reality. Similarly, the driving angular velocity of the wheels cannot be instantly changed. To resolve this issue, the controller's steering output is forced to be pseudo-continuous by limiting the change between time steps. Secondly, when the controller decides to suddenly change the steering angles and apply a certain angular velocity to the wheels, before the new steering angle is applied, the wheels may start to turn. This causes an unintended action to be executed. In order to overcome this problem, the angular velocities are significantly reduced when a sudden steering angle change is detected.

3.1 State Control

While using LQR directly on the states (with unit output matrix) allows one to move the vehicle from a given position and orientation to another, the priorities cannot be properly defined for the controller. An example path is shown in Fig. 4.

As the example suggests, the vehicle is controlled from the origin to the desired arbitrary position, $(0.3, 0.07)$ m, successfully. However, the optimum decision of the LQR does not enable one to control the path. The $\mathbf{Q}$ matrix of the LQR for this example is given in (8).

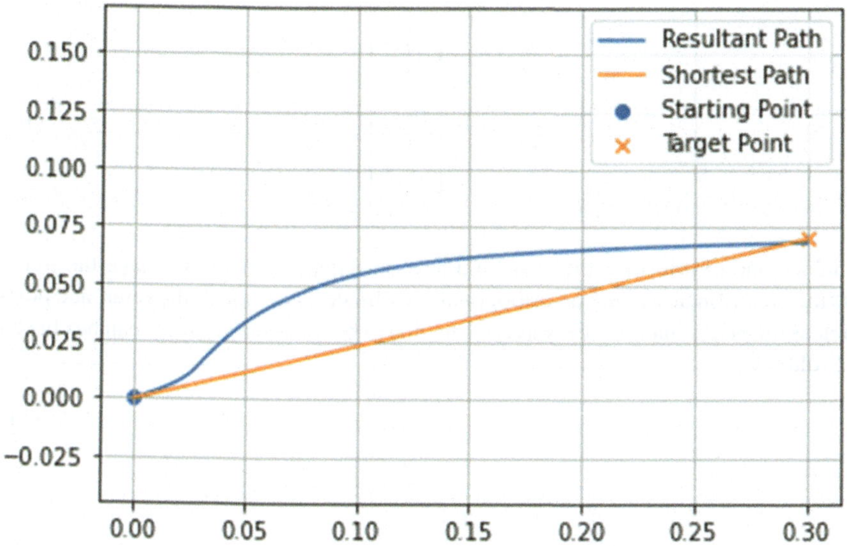

Fig. 4 The resultant path of controlling with unit output matrix in a mathematical simulation

$$\mathbf{Q} = \begin{bmatrix} 1\,0\,0\,0\,0\,0 \\ 0\,1\,0\,0\,0\,0 \\ 0\,0\,1\,0\,0\,0 \\ 0\,0\,0\,0\,0\,0 \\ 0\,0\,0\,0\,0\,0 \\ 0\,0\,0\,0\,0\,0 \end{bmatrix} \tag{8}$$

3.2 *Output Control*

To overcome the issue of not being able to control the path, a suitable output matrix can be defined. The purpose is to define a line that connects the initial and reference positions of the vehicle and use one of the states as the deviation from this line. The same transformation can be applied to the velocities.

Let the line that connects the initial position and the reference position be defined by the equation $ax + by = c$. Choosing the output matrix and reference state as

$$\mathbf{C} = \begin{bmatrix} a & b & 0 & 0 & 0 & 0 \\ a' & b' & 0 & 0 & 0 & 0 \\ 0 & 0 & 1 & 0 & 0 & 0 \\ 0 & 0 & 0 & a & b & 0 \\ 0 & 0 & 0 & a' & b' & 0 \\ 0 & 0 & 0 & 0 & 0 & 1 \end{bmatrix} \tag{9}$$

$$\mathbf{r} = \begin{bmatrix} c \\ c' \\ \theta_{ref} \\ 0 \\ c_{vel} \\ \omega_{ref} \end{bmatrix} \tag{10}$$

enables one to prioritize the error in a new coordinate system. To make the transformed coordinate system an orthogonal coordinate system with the reference position as the origin and the new y axis as the line to be followed, a', b', c' can be chosen as follows.

$$a' = b \tag{11}$$

$$b' = -a \tag{12}$$

$$c' = bx_r - ay_r \tag{13}$$

Let the goal position be $\vec{p}_r = x_r\hat{\imath} + y_r\hat{\jmath}$, and the initial position be $\vec{p}_i = x_i\hat{\imath} + y_i\hat{\jmath}$. The desired velocity can be calculated using the expression

$$c_{vel} = b\vec{g} \cdot \hat{\imath} - a\vec{g} \cdot \hat{\jmath} \tag{14}$$

where $\vec{g}$ is the unit vector in the direction of the desired movement.

$$\vec{g} = (\vec{p}_r - \vec{p}_i)/|\vec{p}_r - \vec{p}_i| \tag{15}$$

Using the described output matrix, the first two error components represent e_1 and e_2 in Fig. 5.

Using the output matrix $\mathbf{C}$ and reference input $\mathbf{r}$, the path given in Fig. 6 can be obtained.

The vehicle is controlled between the same initial and final positions. However, in this example, the errors are properly prioritized to control the vehicle on a line. The $\mathbf{Q}$ matrix of the LQR for this example is given in (16).

$$\mathbf{Q} = \begin{bmatrix} 10 & 0 & 0 & 0 & 0 & 0 \\ 0 & 0.1 & 0 & 0 & 0 & 0 \\ 0 & 0 & 1 & 0 & 0 & 0 \\ 0 & 0 & 0 & 1 & 0 & 0 \\ 0 & 0 & 0 & 0 & 0.1 & 0 \\ 0 & 0 & 0 & 0 & 0 & 0 \end{bmatrix} \tag{16}$$

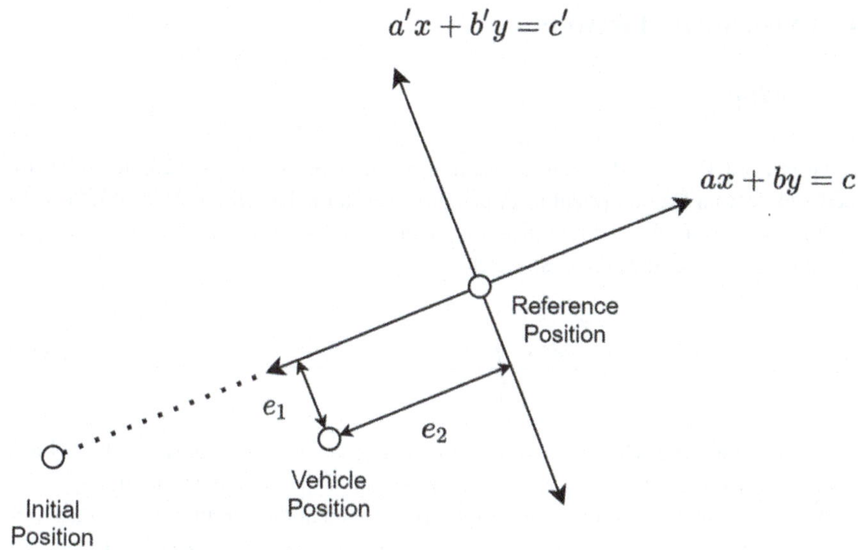

Fig. 5 The transformed coordinate system and the error components

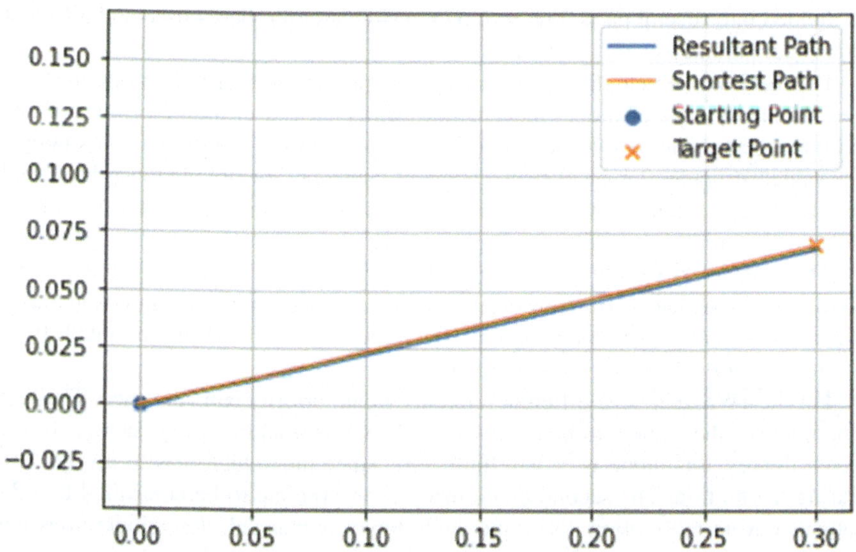

Fig. 6 The resultant path of controlling with the described output matrix in a mathematical simulation

4 Experimental Results

4.1 Setup

The proposed 4WS, 4WD vehicle structure is tested on a prototype vehicle with state feedback from a Vicon optical motion capture system. The structure is compared to a commonly used 2WS vehicle structure in terms of trajectory tracking performance. Equation (17) describes the cost metric.

$$C = \int_{0}^{t_f} |\vec{x}_t(t) - \vec{x}_v(t)|\, dt \tag{17}$$

where t_f is the time when the end point is reached, $\vec{x}_t(t)$ is the position the vehicle has to be on at the time t and $\vec{x}_v(t)$ is the position of the vehicle at the time t.

As a test case, the vehicles are initially positioned at the origin with 0° angle (i.e. facing east). Then, the vehicles are controlled to the position (0, 0.4) m targeting approximately 3.33 cm/s in the direction of the target.

Since the control of a 2WS vehicle requires long term planning, the movement is divided into several parts. For each part, a similar controller is used and different controller parameters are selected.

For the 2WS vehicle, three long term strategies are used. The first one involves the vehicle to make a smooth movement towards the end of the trajectory but this requires the vehicle to considerably deviate from the line. For the second strategy, the vehicle executes 2 maneuvers to change its angle before moving towards the end point. The last strategy involves 4 maneuvers however, this causes the vehicle to delay and increases the trajectory tracking cost.

The maneuvers simply involve the vehicle being controlled to a specific angle until a certain deviation from the desired path is reached. This corresponds to a case where the vehicle has to move in a narrow hallway since deviating too much from the middle of the hallway is not possible.

For the 4WS vehicle, 2 different reference input, priority pairs are used. The first one involves the vehicle to be controlled at 0° while tracking the trajectory. In the given figures, this mode is called "Full Slip" since the vehicle has to always slip during the motion. The second mode involves the vehicle to be controlled to 1.37 radians during the motion. Once the angle target is reached, the vehicle does not have to slip for the rest of the motion. Since movement towards the target without slippage can be achieved in a relatively wide angle range, the angle priority of the controller is selected to be somehow lower compared to the first mode.

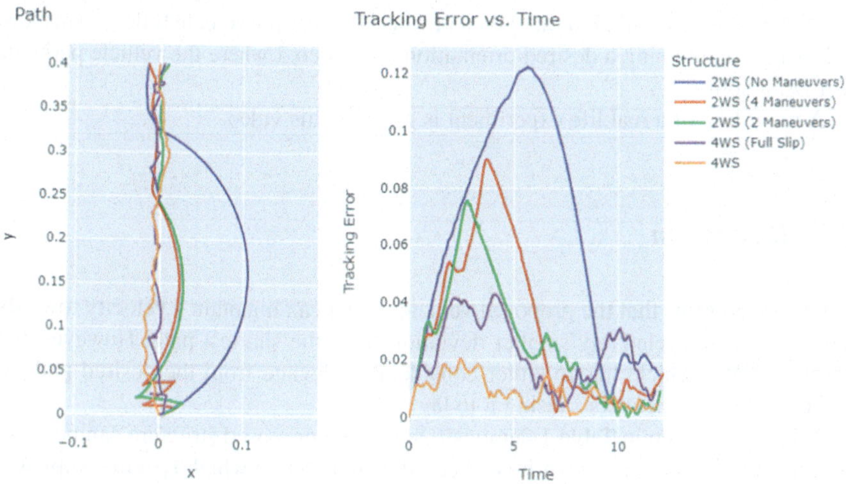

(a) Paths that are followed by the vehicles

(b) Trajectory errors of the vehicles through time

Fig. 7 Paths and trajectory errors

4.2 Results

The paths that are followed by the vehicles and the deviation from the trajectory in the best trial for each are given in Fig. 7a, b respectively.

Using (17) for the given trials, the costs of different structures and modes are given in Table 1. This result shows that making more than 2 maneuvers only causes the 2WS vehicle to delay further. This means that it is not necessary to increase the number of maneuvers any further.

The trajectory tracking errors defined by the metric given in (17) are given in Table 1.

The maximum error in terms of the angle was less then 12° when the 4WS vehicle was controlled to 0° during the motion.

An animation of the movements of different structures in the experiment are given in this video.

Table 1 Costs of different structures and modes

Mode	Cost
2WS (No maneuvers)	0.767
2WS (4 maneuvers)	0.386
2WS (2 maneuvers)	0.319
4WS (Full slip)	0.260
4WS	0.103

Two videos of simulations are given in video 1 where the vehicle follows a narrow hall while maintaining a desired orientation and video 2 where the vehicle parks in a difficult spot.

A recording of a real life experiment is given in this video.

4.3 Discussion

Figure 7a suggests that the proposed vehicle model can maintain a velocity towards the target with a relatively smaller deviation from the shortest path. However, the baseline 2WS vehicle has to either considerably deviate from the desired path or make multiple maneuvers causing it to lag.

The results given in Table 1 demonstrate that the proposed structure can perform significantly better even when it is forced to stay at angle which requires slippage. When the angle requirement is relaxed, a further significant improvement can be obtained for the proposed structure.

5 Conclusion

In this work, it is shown that slippage at low speed can be advantageous while tracking a challenging trajectory. A warehouse robot which has to make sharp maneuvers in a crowded space or a regular car which is trying to parallel park in a tight spot may need to follow a challenging trajectory which may might be possible to follow without slippage.

It is shown that using the proposed structure and the controller, the vehicle can be forced to slip in a desired direction without prior maneuvers. It is also important to note that the proposed structure can avoid slippage unless it is necessary. Therefore the adverse effects of excessive slippage can be avoided when it is not necessary to slip.

Even though a similar amount of maneuverability may be reached by complex wheel structures such as omni wheels, the proposed method demonstrates that it can be done with simpler wheels.

Acknowledgements METU-ROMER supported this work by providing access several to manufacturing methods, electronic components and the motion tracking system Vicon. This work would not have been possible without the support of METU-ROMER.

References

1. Deepak SV, Naresh C, Hussain SA (2012) Modelling and analysis of alloy wheel for four wheeler vehicle. Int J Mech Eng Robot Res 1(3):72–80
2. Choudhari DS (2014) Four wheel steering system for future. Int J Mech Eng Robot Res 3(4):383–387
3. Saxena S, Kumar V, Luthra SS, Kumar A (2014) Wheel steering systems (4WAS). Int J Mech Eng Robot Res Special Issue 1(1):213–218
4. Rajasekar R, Pranavkarthik KP, Prashanth R, Kumar SS (2013) Design and fabrication of staircase climbing wheelchair. Int J Mech Eng Robot Res 2(2):320–323
5. Jhino S, Lincol V, Josue L, Junior Q, Carlos M (2018) Maneuverability study of a vehicle with rear wheel steering. Mech. Mach. Sci. 54:559–567. https://doi.org/10.1007/978-3-319-67567-1_52/FIGURES/7
6. Mondal K, Wallace B, Rodriguez AA (2020) Stability versus maneuverability of non-holonomic differential drive mobile robot: focus on aggressive position control applications. In: CCTA 2020—4th IEEE conference on control technology and applications, pp 388–395. https://doi.org/10.1109/CCTA41146.2020.9206155
7. Xu G, Wang D (2006) Full state tracking of a four-wheel-steering vehicle based on output tracking control strategies. In: 9th international conference on control, automation, robotics vision, 2006, ICARCV '06. https://doi.org/10.1109/ICARCV.2006.345154
8. Kang J, Yoo J, Yi K (2011) Driving control algorithm for maneuverability, lateral stability, and rollover prevention of 4WD electric vehicles with independently driven front and rear wheels. IEEE Trans. Veh. Technol. 60(7):2987–3001. https://doi.org/10.1109/TVT.2011.2155105
9. Bayar G, Bergerman M, Konukseven EI, Koku AB (2016) Improving the trajectory tracking performance of autonomous orchard vehicles using wheel slip compensation. Biosyst Eng 146:149–164. https://doi.org/10.1016/J.BIOSYSTEMSENG.2015.12.019
10. Hafizah N et al (2016) Modelling and control strategies in path tracking control for autonomous ground vehicles: a review of state of the art and challenges. J Intell Robot Syst 86(2):225–254. https://doi.org/10.1007/S10846-016-0442-0
11. Taghia J, Wang X, Lam S, Katupitiya J (2017) A sliding mode controller with a nonlinear disturbance observer for a farm vehicle operating in the presence of wheel slip. Auton. Robot. 41(1):71–88. https://doi.org/10.1007/S10514-015-9530-4/TABLES/4
12. Yin GD, Chen N, Wang JX, Chen JS (2010) Robust control for 4WS vehicles considering a varying tire-road friction coefficient. Int J Automot Technol 11(1):33–40. https://doi.org/10.1007/S12239-010-0005-5
13. Pilisiewicz J, Kaczyński R (2017) Geometric analysis of maneuverability performance for vehicles with two steering axles. Transp Probl 12(2):43–52. https://doi.org/10.20858/tp.2017.12.2.5
14. Choi MW, Park JS, Lee BS, Lee MH (2008) The performance of independent wheels steering vehicle(4WS) applied Ackerman geometry. In: 2008 International Conference on Control, Automation, Systems ICCAS, 2008, pp 197–202. https://doi.org/10.1109/ICCAS.2008.4694549
15. Bayar G, Koku AB, Konukseven EI (2015) Dynamic modeling and parameter estimation for traction, rolling, and lateral wheel forces to enhance mobile robot trajectory tracking. Robotica 33(10):2204–2220. https://doi.org/10.1017/S0263574714001386
16. Kane T, Levinson D (1985) Dynamics, theory and applications. McGraw-Hill, New York

Design and Construction of a Prototype for the Lyophilization Process

Erick Coronel M., Jessenia López O., Patricia Constante P., Cristina Sánchez L., Andres Ortega C., Andrea Tobar, Jines David E., and Andrea Lescano

Abstract This study describes the design, construction and prototype control of a lyophilizer based on parameters of freezing temperature, heating and vacuum pressure, seeking the conservation of organoleptic properties of fruits and vegetables in the region. The prototype is designed with a drying chamber insulated with expanded polyurethane and is hermetically sealed with an ap-proximate capacity of 10 lbm of fresh product distributed in 3 AISI 304 1 mm stainless steel trays. The prototype consists of 3 subsystems such as: the freezing subsystem that works at temperatures below −35 °C, the heating sub-system that adds energy in the form of heat to the food, raising its temperature to 50 °C, and the vacuum subsystem in which it is generates a pressure of −24 in Hg. Finally, the organoleptic properties of the freeze-dried beet were evaluated by duo trio analytical sensory tests and magnitude estimation, concluding that the color, flavor and smell were not modified; however, the texture of the fruit was altered, acquiring a crunchy composition.

E. Coronel M. · J. López O. (✉) · P. Constante P. · A. Ortega C. · A. Tobar
Universidad de las Fuerzas Armadas ESPE-L, Latacunga, Ecuador
e-mail: jelopez14@espe.edu.ec

E. Coronel M.
e-mail: eacoronel@espe.edu.ec

P. Constante P.
e-mail: pnconstante@espe.edu.ec

A. Ortega C.
e-mail: asortega3@espe.edu.ec

A. Tobar
e-mail: aytovar@espe.edu.ec

C. Sánchez L. · J. David E. · A. Lescano
Instituto Superior Tecnológico Tungurahua, Ambato, Ecuador
e-mail: cdsanchez3@espe.edu.ec

J. David E.
e-mail: djines@institutos.gob.ec

A. Lescano
e-mail: alescano@institutos.gob.ec

G. L. Conte and O. Sename (eds.), *Proceedings of the 11th International Conference on Mechatronics and Control Engineering*, Lecture Notes in Mechanical Engineering,
https://doi.org/10.1007/978-981-99-6523-6_13

Keywords Prototype · Lyophilization · Design · Control

1 Introduction

Ecuador, being a tropical country, produces various fruits and vegetables; therefore, a method is sought to preserve its quality for long periods of time; due to its rapid deterioration, it is difficult to store the products. One of the methods used to solve this problem is freezing, but this can damage the organoleptic properties of the food due to the formation of crystals inside the membranes [1].

The use of new technologies is the right way to preserve food and one of them is freeze-drying; for this reason, this research describes the design of a dehydrator prototype with up to 3 trays of 1250 cm^2 area for fruit minced, which consists of the phases of freezing, primary drying and secondary drying inside a hermetic cabin that is subjected to vacuum in the second and third stages of the process; In addition, the prototype has an automatic work cycle.

2 Prototype Design

Lyophilization is a dehydration process by sublimation [2]. The process consists of freezing the product at very low temperatures, eliminating water to reach humidity levels of less than 5% of its initial content [3]. Initially the process this at environmental conditions, then the temperature is reduced to change the liquid to solid state, later is to freeze the food and when the pressure inside the chamber decreases and is reached to sublimation ice into vapor below the triple point of water; finally, the vacuum be maintained and the temperature of the food raised, which provides it with the energy necessary for the sublimation of the greater water amount [4]. For the design of the prototype, the freezing of the food below its freezing point was considered approximately −40 °C, then the pressure is lowered below the triple point of the water and creates a vacuum with a temperature of −35 to −40 °C; finally, is maintains the generated vacuum and raises the temperature of the product to 60 °C to remove the partially bound water of the product.

2.1 Mechanical Design

For the mechanical design of the prototype a rectangular and simple chamber configuration was selected (see Fig. 1). For respect the sanitary and hygienic standards in the environment, materials must be selected that do not rust and that withstand the external load within the allowed range of deformation, without altering the workability of the machine [5], so AISI 304 1 mm stainless steel was selected, then an analysis was carried out to verify the safety of the design, as can be seen in Table 1, the values of deformations and stresses that were generated inside the chamber do not exceed the established limits. Once the structure was designed, it was analyzed by CAE with finite element analysis with which the displacements, unitary deformations and stresses in the structure (see Table 1).

The factor of safety for average materials operating in ordinary environments, subjected to loads and stresses that can be determined with relative precision should be 2–2.5 [6]; based on the analysis, the Von Mises stress was obtained and the safety coefficient was calculated at the point of maximum stress of 6.5 indicates great safety in the face of failure.

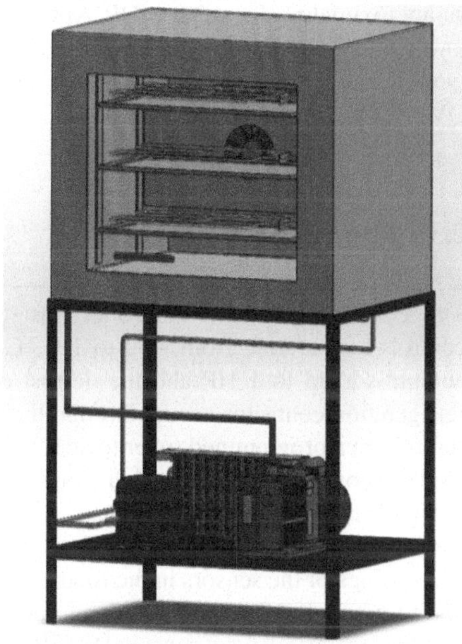

Fig. 1 Prototype in CAD software

Table 1 Von Mises stresses, displacements and unit strains

Name	Type	Mín	Máx
Stresses1	VON: Von Mises stress	2.883e + 03 N/m^2 Node: 14865	6.379e + 07 N/m^2 Node: 4957
Displacments1	URES: Resulting Displacements	0.000e + 00 mm Node: 5493	4.755e−01 mm Node: 5527
Unit strains1	ESTRN: Equivalent Unit strain	5.884e−07 Item: 2034	6.810e−04 Item: 3089

2.2 Sizing of the Freezing System

For the calculation of the thermal load of the product, the data of the average latent heat, specific heats and speed factor of "Heat and mass transfer fundamentals and applications" book [7], giving a cooling heat value of 456.327 [Btu], the latent heat of freezing point of 1097.67 [Btu] and the sensible heat below freezing point of 306.56 [Btu]. The cooling of the food inside the lyophilization chamber is carried out by circulating the cold air through a small fan that generates a power consumption of 27.5 [W]. The total sum of all the loads inside the lyophilization chamber is of 0.26 HP, the real value suffers losses due to factors unrelated to the design such as compressor efficiency or leaks, for which we add 30% of safety power, obtaining an approximate value of 0.35 HP. After the freezing process, the pressure must be lowered to −610 Pa in the center of the chamber; for this reason, the volume of the lyophilization chamber was calculated, obtaining a value of 4.65 ft^3, for which the QVP-900 pump was selected.

2.3 Control System

The TC-900E electronic controller was selected for the refrigeration system, for controls temperature from −50 to 105 °C. The relay out-put directly controls the compressor up to 1 HP and the defrost output is up to 10 A. Activation of the refrigeration contactor occurs through the control relay output and is connected in series with a programmed timer to add freeze and preheat times. After this time, the chiller stops working for the final stage of production. When the freezing phase is complete to the temperature is reached, the vacuum system and the heating system are activated. The heating system activates or deactivates the resistances based on the readings of the sensors in the food at the desired temperature during the primary drying process. An Arduino Uno microcontroller was chosen to control these two systems, we also have a sensor (DS18B20) in contact with the food as input and a resistor and a vacuum circuit as output. For chamber cooling unit, the refrigeration unit through its TC-900 E controller drives the compressor nominal load amperage of 9.6 A with a voltage of 115 V and the condenser fan of 0.48 A with a voltage of 115 V at 60 Hz. With which the current that the conductor obtaining a value a current of the condenser of 12.48 A and of the fan of 24.48 A.

After sizing the electrical components, a force and control diagram was made in which we can see all the components (Fig. 2).

In the Fig. 3 shows the connection of the two solenoid valves considered, they will operate in parallel with each vacuum pump, the connection of the electrical control is also detailed of the refrigeration unit in series with a timer which allows us to deactivate the refrigeration system. The control scheme also includes a two-position ignition selector, the display shows the temperature of the chamber and the time that

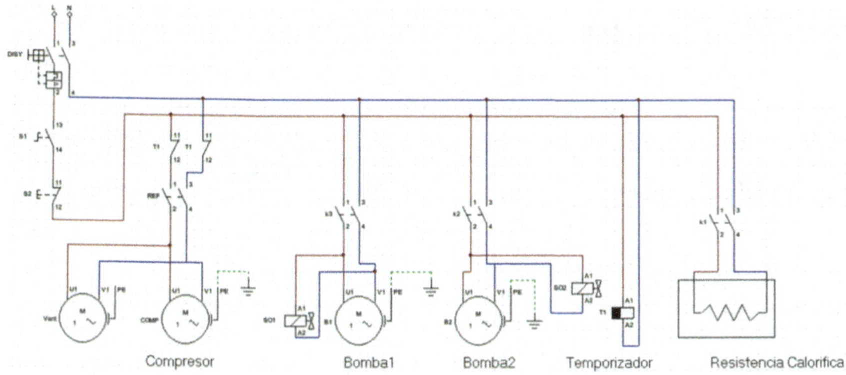

Fig. 2 Power circuit diagram

has elapsed in the process, an emergency stop button, the controller also activates resistance contactors, switching relays for pump 1 and 2.

For the power system in Fig. 4, the outputs of the Arduino Uno microcontroller are observed, they activate the coils with high or low pulses at 5 V dc of the relays, which allow the passage of the phase line at 110 V towards coil A1 of the contactors on the power side.

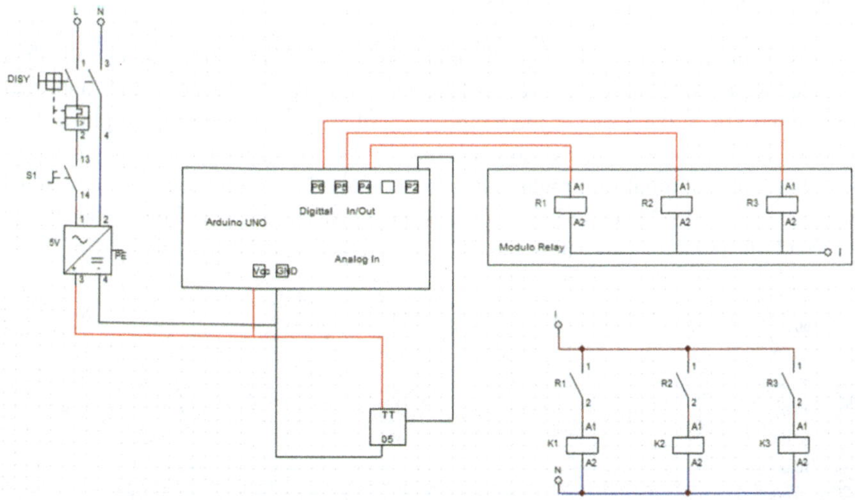

Fig. 3 Control circuit diagram

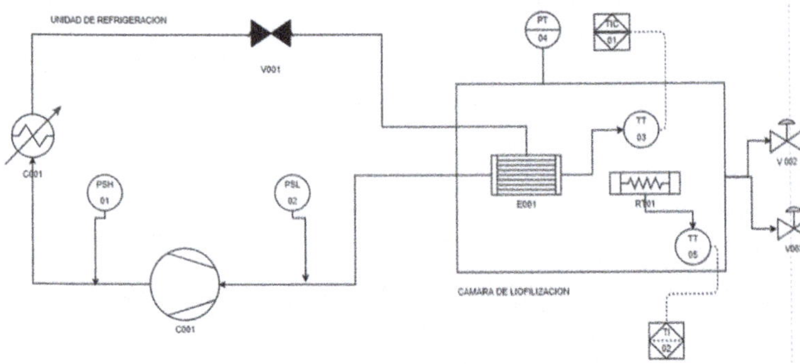

Fig. 4 P&ID piping and instrumentation diagram

3 Testing and Analysis of Results

3.1 Testing the System as a Whole

After calibrating the sensors and checking the operation of each subsystem, the prototype was tested in vacuum to verify the correct operation of the code implemented in the microcontroller, a screen is displayed to indicate whether the processes it is in freezing, pre-drying or secondary drying. The temperature measurement was recorded in an Excel file through the serial port of the controller and the vacuum level as a function of time through the analog reading of the vacuum gauge, the data obtained were plotted and it was verified that they comply with the curve of lyophilization. In the Fig. 5, the curve of the temperature of the food during the lyophilization process can be observed, the freezing reaches −35 °C (blue) and passes to the primary desiccation or sublimation in which heat is added and its temperature is maintained between −20 and −25 °C (red), in the secondary desiccation the refrigeration system is turned off and the temperature is raised between 45 and 50 °C (green) (Fig. 5).

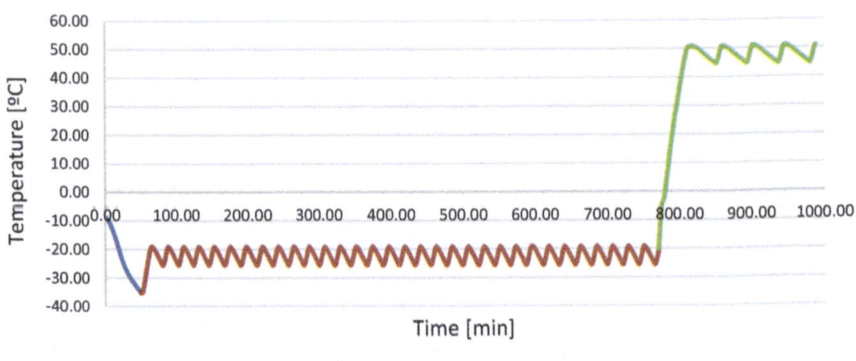

Fig. 5 Temperature versus real time of the process

3.2 Organoleptic Tests

Sensory tests were carried out with 10 judges to verify the conservation of organoleptic properties of freeze-dried beets (color, smell, flavor and texture), each judge was assigned a reference sample of fresh fruit and a freeze-dried fruit [8]. Once all the tests were finished, the ANOVA test was tabulated.

According to the results of Table 2, 3 and 4 we can observe that the critical value for F is greater than the value of F calculated, for which the null hypothesis (H_0) is not rejected, another factor to analyze is the value of P (probability), which confirms the hypothesis that the freeze-dried sample is the same as the fresh sample in terms of taste, smell and color; however, in Table 5 we can see that the critical value for F is less than the value of F calculated, for which the null hypothesis (H_0) is rejected. Another factor to analyze is the value of P (probability); therefore, the lyophilized sample is not the same as the fresh sample in terms of texture, this is due to process.

Table 2 Analysis of variance of flavor

Origin of variations	Average of the squares	F	Probability	Critical value for F
Between groups	0.002631281	0.2397	0.7884	2.5106
Within groups	0.010973524			
Total				

Table 3 Analysis of variance of odor

Origin of variations	Average of the squares	F	Probability	Critical value for F
Between groups	0.004581604	0.6355	0.53737	3.3541
Within groups	0.0072086			
Total				

Table 4 Analysis of variance of color

Origin of variations	Average of the squares	F	Probability	Critical value for F
Between groups	0.00951	3.36721	0.04947	4.004
Within groups	0.00282			
Total				

Table 5 Analysis of variance of texture

Origin of variations	Average of the squares	F	Probability	Critical value for F
Between groups	0.0660	9.42475	0.000785	3.354130
Within groups	0.0070			
Total				

4 Conclusions

The modeling of the prototype was carried out in a CAD software to establish the dimensions chamber of 60 cm wide, 50 cm high, 50 cm of deep and 3 trays with capacity of 10 lbm of fresh product. The design analysis using CAE verify that the prototype is safe; therefore, a dehydration system was designed of lyophilization, with three stages: freezing to −35 °C and atmospheric pressure, primary desiccation at −10 °C and a vacuum pressure of less than 610 Pa must be maintained and secondary desiccation to 60 °C at a vacuum pressure of less than 610 Pa.

The electrical system was dimensioned, a Full Gauge TC-900E temperature controller was used, an Arduino Uno microcontroller was used for the vacuum and heating subsystems according to the programming for each of the freeze- drying stages. On the other hand, the result of the lyophilization process was evaluated by duo-trio analytical sensory tests and by estimation of magnitude with 10 judges and with three samples to evaluate one of fresh product and two of lyophilized product, in the beet the flavor, smell and color while the texture did change.

5 Future Works

For future projects, it is recommended to implement a steam trap, also called a condensation chamber, which prevents damage to the vacuum pump and prevents the product from being rehydrated by the circulation of steam and condensation inside the work chamber. In addition, the implementation of an HMI (Human Machine Interface) for the management of recipes and easy change of parameters should be taken as a new research work for the improvement of the prototype.

References

1. Garcìa C (2018) «Liofilizaciòn de frutillas enteras: efecto de micro-perforaciones realizadas con tecnologìa laser de CO2 en el tiempo de secado primario,» Universidad Tècnica Federico Santa Marìa, Valparaìso-Chile
2. Orrego C (2003) Procesamiento de alimentos. Centro de Publicaciones Universidad Nacional de Colombia Sede Manizales, Manizales

3. Orrego C (2008) Congelaciòn y liofilizaciòn de alimentos. Artes Gráficas Tizan Ltda, Manizales
4. Falconi J, Valdiviezo y C, Ramirez L (2021) «Prediction of the freeze-drying time of Araza (Eugenia Stipitata) using mathematical models» Ecuadorian Sci J 5(4):89–97
5. Lòpez P y, Padilla D (2012) «Diseño y construcciòn de un equipo hidrosolubilizador automatico de extractos botànicos para el manejo agronomico por medio de liofilizaciòn para el laboratorio de fitoquimica del IASA 1» UFA ESPE, Sangolqui
6. Arteaga O (2018) Resumen de la asignatura de diseño de elementos de máquina, UFA-ESPE, Latacunga, pp 1–139
7. Cengel Y y, Ghajar A (2011) Heat and mass transfer fundamentals and applications. Mexico: McGraw-Hill/interamericana editores, S.A. DE C.V
8. Espinosa J (2007) Evaluaciòn sensorial de los alimentos. Editorial Universitaria, Cuba

The manufacturer's authorised representative in the EU is Springer Nature Customer Service Centre GmbH, Europaplatz 3, 69115 Heidelberg, Germany. If you have any concerns regarding our products, please contact ProductSafety@springernature.com

Printed and bound by CPI Group (UK) Ltd, Croydon, CR0 4YY
30/09/2026
02225301-0003